Physical Geography

Physical Geography

PHYSICAL GEOGRAPHY

Girish Chopra

Commonwealth

Physical Geography

Edition : 2018

ISBN 81-7169-999-5

Published by :
COMMONWEALTH PUBLISHERS
4831/24, Prahlad Street,
Ansari Road, Darya Ganj,
New Delhi - 110002
Ph. : 23275511, 23257835
e-mail : commonwealthbooks@gmail.com

Typesetting by :
Rishu Computers
Delhi

Printed at :
Roshan Offset Printers
Delhi

Preface

Geography is the scientific study of the locational and spatial variation in both physical and human phenomena on Earth. The word derives from the Greek words *ge* (the Earth) and *graphein* (to write, as in to describe). Geography is much more than cartography, the study of maps. It not only investigates what is where on the Earth, but also why it's there and not somewhere else, sometimes referred to as 'location in space'. It studies this whether the cause is natural or human. It also studies the consequences of those differences.

Physical geography is the study natural and human constructed phenomena relative to a spatial dimension. It focuses on geography as an earth science, making use of biology to understand global flora and fauna patterns, and mathematics and physics to understand the motion of the earth and relationship with other bodies in the solar system. It also includes landscape ecology and environmental geography.

The purpose of this book is to provide the reader with an overview of the elements that constitute the physical environment. It provides an introduction to the major aspects of physical geography: the energy-atmosphere system; the water, weather, and climate systems; the earth-atmosphere interface; and, soil, ecosystems, and biomes.

It stresses the interrelationships between environmental systems and worldwide distributions of phenomena. The importance of the physical environment to humans are also emphasized.

Offering a fully integrated, systems-based approach to physical geography, this book incorporates relevant and recent developments in the field. Its' accurate and comprehensive coverage provides both the breadth and depth necessary to appreciate how humans are changing, and are changed by, the Earth. The book is ideally suited to all who concerned with geography and earth science.

Girish Chopra

Contents

1

Physical Geography

Physical geography is a discipline that is part of a much larger area of understanding called Geography. Most individuals define geography as a field of study that deals with maps. This definition is only partially correct. A better definition of geography may be the study natural and human constructed phenomena relative to a spatial dimension. The discipline of geography has a history that stretches over many centuries. Over this time period, the study of geography has evolved and developed into an important form of human scholarship. Examining the historical evolution of geography as a discipline provides some important insights concerning its character and methodology. These insights are also helpful in gaining a better understanding of the nature of physical geography.

History of Physical Geography

The nature of understanding in Physical Geography has changed over time. When investigating this change it becomes apparent that certain universal ideas or forces had very important ramifications to the academic study of Physical Geography. During the period from 1850 to 1950, there seems to be four main ideas that had a strong influenced on the discipline:

(1) *Uniformitarianism* : this theory rejected the idea that catastrophic forces were responsible for the current conditions on the Earth. It suggested instead that continuing uniformity of existing processes were responsible for the present and past conditions of this planet.

(2) *Evolution* : Charles Darwin's Origin of Species suggested that natural selection determined which individuals would pass on their genetic traits to future generations. As a result of this theory, evolutionary explanations for a variety of natural phenomena were postulated by scientists. The theories of uniformitarianism and evolution arose from a fundamental change in the way humans explained the universe and nature. During the 16th, 17th, and 18th centuries scholars began refuting belief or myth based explanations of the cosmos, and instead used science to help explain the mysteries of nature. Belief based explanations of the cosmos are made consistent with a larger framework of knowledge that focuses on some myth. However, theories based on science questioned the accuracy of these beliefs.

(3) *Exploration and Survey* : much of the world had not been explored before 1900. Thus, during this period all of the fields of Physical Geography were actively involved with basic data collection. This data collection included activities like determining the elevation of land surfaces, classification and description of landforms, the measurement of the volume of flow of rivers, measurement of various phenomena associated to weather and climate, and the classification of soils,

(4) *Conservation* : beginning in the 1850s a concern for the environment began to develop as a result of the human development of once natural areas in the United States and Europe. One of the earliest statements of these ideas came from George Perkins Marsh in his book "Man in Nature" or "Physical Geography as Modified by Human Action". This book is often cited by scholars as the first significant academic contribution to conservation and environmentalism.

After 1950, the following two forces largely determined the nature of Physical Geography:

(1) *The Quantitative Revolution* : measurement became the central focus of research in Physical Geography. It was used primarily for hypothesis testing. With measurement came mapping, models, statistics, mathematics, and hypothesis testing. The quantitative revolution was also associated with a change in the way in which physical geographers studied the Earth and its phenomena. Researchers now began investigating process rather than mere description of the environment.

(2) *The study of Human/Land Relationships* : the influence of human activity on the environment was becoming very apparent after 1950. As a result, many researchers in Physical Geography began studying the influence of humans on the environment. Some of the dominant themes in these studies included: environmental degradation and resource use; natural hazards and impact assessment; and the effect of urbanization and land-use change on natural environments.

Scope of Physical Geography

Physical Geography also examines the interrelationships of these phenomena to human activities. This sub-field of Geography is academically known as the Human-Land Tradition. This area of Geography has seen very keen interest and growth in the last few decades because of the acceleration of human induced environmental degradation. Thus, Physical Geography's scope is much broader than the simple spatial study of nature. It also involves the investigation of how humans are influencing nature. Academics studying Physical Geography and other related Earth Sciences are rarely generalists.

Most are in fact highly specialized in their fields of knowledge and tend to focus themselves in one of the following well defined areas of understanding in Physical Geography:

— *Geomorphology* : studies the various landforms on the Earth's surface.

— *Pedology* : is concerned with the study of soils.

— *Biogeography* : is the science that investigates the spatial relationships of plants and animals.

— *Hydrology* : is interested in the study of water in all its forms.

— *Meteorology* : studies the circulation of the atmosphere over short time spans.

— *Climatology* : studies the effects of weather on life and examines the circulation of the atmosphere over longer time spans.

The above fields of knowledge generally have a primary role in introductory textbooks dealing with Physical

Geography. Introductory Physical Geography textbooks can also contain information from other related disciplines including:

— *Geology* : studies the form of the Earth's surface and subsurface, and the processes that create and modify it.

— *Ecology* : the scientific study of the interactions between organisms and their environment.

— *Oceanography* : the science that examines the biology, chemistry, physics, and geology of oceans.

— *Cartography* : the technique of making maps.

— *Astronomy* : the science that examines celestial bodies and the cosmos.

Future of Physical Geography

The following list describes some of the important future trends in Physical Geography research:

(1) Continued development of Applied Physical Geography for the analysis and correction of human-induced environmental problems. A student of Applied Physical Geography uses theoretical information from the field of Physical Geography to manage and solve problems related to natural phenomena found in the real world.

(2) *Remote Sensing* : Advances in technology have caused the development of many new instruments for the monitoring of the Earth's resources and environment from airborne and space platforms (see three-dimensional image of hurricane Andrew, Landsat image of San Francisco Bay, Landsat image of Vancouver, British Columbia, and a space radar image

of Victoria, British Columbia). The most familiar use of remote sensing technology is to monitor the Earth's weather for forecasting.

(3) *Geographic Information Systems* : A geographic information system (GIS) merges information in a computer database with spatial coordinates on a digital map. Geographic Information Systems are becoming increasingly more important for the management of resources.

Scientific Method

Francis Bacon, a 17th century English philosopher, was the first individual to suggest a universal methodology for science. Bacon believed that scientific method required an inductive process of inquiry. Karl Popper later refuted this idea in the 20th century. Popper suggested that science could only be done using a deductive methodology.

Science is simply a way of acquiring knowledge about nature and the Universe. To do science, one must follow a specific universal methodology. The central theme of this methodology is the testing of hypotheses. The overall goal of science is to better understand the world around us. Various fields of study, like Physics, Chemistry, Biology, Medicine and the Earth Sciences, have used science exclusively to expand their knowledge base. Science allows its practitioners to acquire knowledge using techniques that are both neutral and unbiased.

The broadest, most inclusive goal of science is to understand (Figure 1). Understanding encompasses a number of other goals of science, many of which are quite specialized (Figure 2). Explanation is perhaps the most important basic goal of understanding. Explanation consists of relating observed reality to a system of concepts, laws, or empirically based generalizations. Explanation may also relate observed phenomena to a system of causes, or link them to mechanisms that are hierarchically structured at lower-levels of function.

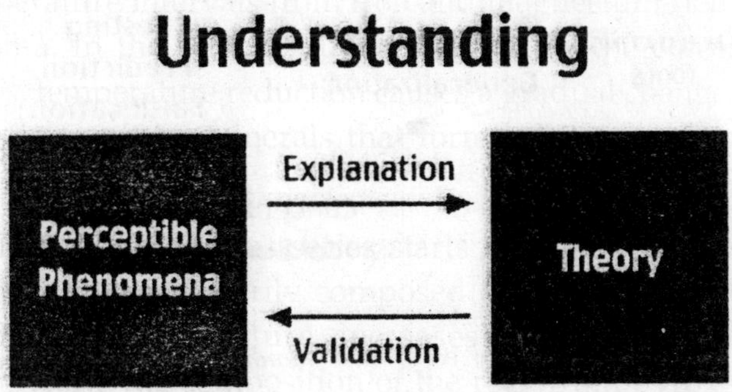

Figure 1. Relationship between phenomena, theory, and understanding using scientific method.

The secondary goal of explanation has two important components: generalization and unification (Figure 2). Generalization may be considered to be the condensation of a body of empirical fact into a simple statement. In the process of such condensation, it is likely that some detail must be omitted and the processes and phenomenon abstracted. Generalization may also involve isolating the phenomenon from other aspects of the system of interest. This is sometimes referred to as idealization. A second aspect of explanation is the unification of apparently unrelated phenomena in the same abstract or ideal system of concepts.

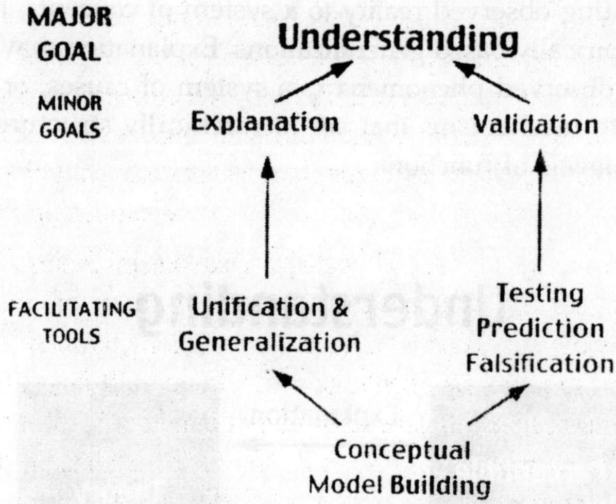

Figure 2. Processes involved in understanding natural phenomena using scientific method.

Another minor goal of science is the validation of constructed models (conceptual model building) of understanding. Validation is accomplished through hypothesis testing, prediction, and falsification.

The Hypothetico-Deductive Method

Philosopher Karl Popper suggested that it is impossible to prove a scientific theory true by means of induction, because no amount of evidence assures us that contrary evidence will not be found. Instead, Karl Popper proposed that proper science is accomplished by deduction. Deduction involves the process of falsification. Falsification is a particular specialized aspect of hypothesis testing. It involves stating some output from theory in specific and then finding contrary cases using experiments or observations. The methodology proposed by Popper is commonly known as the hypothetico-deductive method.

Popper's version of scientific method first begins with the postulation of a hypothesis. A hypothesis is an educated guess or a theory that explains some phenomenon. The researcher then tries to prove or test this scientific theory false through prediction or experimentation (Figure 2). A prediction is a forecast or extrapolation from the current state of the system of interest. Predictions are most useful if they can go beyond simple forecast. An experiment is a controlled investigation designed to evaluate the outcomes of causal manipulations on some system of interest. To get a better understanding of the hypothetico-deductive method, we can examine the following geographic phenomena.

In the brackish tidal marshes of the Pacific Coast of British Columbia and Washington, we find that the plants in these communities spatially arrange themselves in zones that are defined by elevation. Near the shoreline plant communities are dominated primarily by a single species known as Scirpus americanus. At higher elevations on the tidal marsh Scirpus americanus disappears and a species called Carex lyngbyei becomes widespread. The following hypothesis has been postulated to explain this unique phenomenon:

The distribution of Scirpus americanus and Carex lyngbyei is controlled by their tolerances to the frequency of tidal flooding. Scirpus americanus is more tolerant of tidal flooding than Carex lyngbyei and as a result it occupies lower elevations on the tidal marsh. However, Scirpus americanus cannot survive in the zone occupied by Carex lyngbyei because not enough flooding occurs. Likewise, Carex lyngbyei is less tolerant of tidal flooding than Scirpus americanus and as a result it occupies higher elevations on the tidal marsh. Carex lyngbyei cannot survive in the zone occupied by Scirpus americanus because too much flooding occurs.

According to Popper, to test this theory a scientist would now have to prove it false. As discussed above this can be done in two general ways: 1) predictive analysis; or 2) by way of experimental manipulation. Each of these methods has been applied to this problem and the results are described below.

Predictive Analysis

If the theory is correct, we should find that in any tidal marsh plant community that contains Scirpus americanus and Carex lyngbyei that the spatial distribution of these two species should be similar in all cases. This is indeed true. However, there could be some other causal factor, besides flooding frequency, that may be responsible for these unique spatial patterns.

Experimental Manipulation

If the two species are transplanted into the zones of the other they should not be able to survive. An actual transplant experiment found that Scirpus americanus can actually grow in the zone occupied by Carex lyngbyei, while Carex lyngbyei could also grow at lower Scirpus sites. However, this growth became less vigorous as the elevation became lower and at a certain elevation it could not grow at all. These results falsify the postulated theory. So the theory must be modified based on the results and tested again.

The process of testing theories in science is endless. Part of this problem is related to the complexity of nature. Any one phenomenon in nature is influenced by numerous factors each having its particular cause and effect. For this reason, one positive test result is not conclusive proof that the phenomenon under study is explained. However, some tests are better than others and provide us with stronger confirmation. These tests usually allow for the isolation of the phenomena from the effects of causal factors. Manipulative experiments tend to be better than tests based on prediction in this respect.

Concepts of Time and Space in Physical Geography

The concepts of time and space are very important for understanding the function of phenomena in the natural world. Time is important to Physical Geographers because the spatial patterns they study can often only be explained in historic terms. The measurement of time is not absolute. Time is perceived by humans in a relative fashion by using human created units of measurement. Examples of human created units of time are the measurement of seconds, minutes, hours, and days.

Geographers generally conceptualize two types of space. Concrete space represents the real world or environment. Abstract space models reality in a way that distills much of the spatial information contained in the real world. Maps are an excellent example of abstract space. Finally, like time, space is also perceived by humans in a relative fashion by using human created units of measurement.

Both time and space are variable in terms of scale. As such, researchers of natural phenomena must investigate their subjects in the appropriate temporal and/or spatial scales. For example, an investigator studying a forest ecosystem will have to deal with completely different scales of time and space when compared to a researcher examining soil bacteria. The trees that make up a forest generally occupy large tracts of land. For example, the boreal forest occupies millions of hectares in Northern Canada and Eurasia. Temporally, these trees have life spans that can be as long as several hundred years. On the other hand, soil bacteria occupy much smaller spatial areas and have life spans that can be measured in hours and days.

Study of Form or Process?

Physical Geography as a science is experiencing a radical change in philosophy. It is changing from a science that was highly descriptive to one that is increasingly experimental and theoretical. This transition represents a strong desire by Physical Geographers to understand the processes that cause the patterns or forms we see in nature.

Before 1950, the main purpose of research in Physical Geography was the description of the natural phenomena. Much of this description involved measurement for the purpose of gaining basic facts dealing with form or spatial appearance. Out of this research Physical Geographers determined such things as: the climatic characteristics for specific locations and regions of the planet; flow rates of rivers; soil characteristics for various locations on the Earth's surface; distribution ranges of plant and animal species; and calculations of the amount of freshwater stored in lakes, glaciers, rivers and the atmosphere.

By the beginning of the 20th century Physical Geographers began to examine the descriptive data that was collected, and started to ask questions related to why? Why is the climate of urban environments different from the climate of rural? Why does hail only form in thunderstorms? Why are soils of the world's tropical regions nutrient poor? Why do humid and arid regions of the world experience different levels of erosion?

In Physical Geography, and all other sciences, most questions that deal with why are usually queries about process. Some level of understanding about process can be derived from basic descriptive data. Process is best studied, however, through experimental manipulation and hypothesis testing. By 1950, Physical Geographers were more interested in figuring out process than just collecting descriptive facts about the world. This attitude is even more prevalent today because of our growing need to understand how humans are changing the Earth and its environment.

Finally, as mentioned above, a deeper understanding of process normally requires the use of hypothesis testing, experimental methods, and statistics. As a result, the standard undergraduate and graduate curriculum in Physical Geography exposes students to this type of knowledge so they can better ask the question why?

Descriptive Statistics

Physical Geographers often collect quantitative information about natural phenomena to further knowledge in their field of interest. This collected data is then often analyzed statistically to provide the researcher with impartial and enlightening presentation, summary, and interpretation of the phenomena understudy. The most common statistical analysis performed on data involves the determination of descriptive characteristics like measures of central tendency and dispersion.

It usually is difficult to obtain measurements of all the data available in a particular system of interest. For example, it may be important to determine the average atmospheric pressure found in the center of hurricanes. However, to make a definitive conclusion about a hurricane's central pressure with 100 % confidence would require the measuring of all the hurricanes that ever existed on this planet. This type of measurement is called a population parameter. Under normal situations, the determination of population parameters is impossible, and we settle with a subset measure of the population commonly called an estimator.

Estimators are determined by taking a representative sample of the population being studied. Samples are normally taken at random. Random sampling implies that each measurement in the population has an equal chance of being selected as part of the sample. It also ensures that the occurrence of one measurement in a sample in no way influences the selection of another. Sampling methods are biased if the recording of some influences the recording of others or if some members of the population are more likely to be recorded than others.

Measures of Central Tendency

Collecting data to describe some phenomena of nature usually produces large arrays of numbers. Sometimes it is very useful to summarize these large arrays with a single parameter. Researchers often require a summary value that determines the center in a data sample's distribution. In orther words, a measure of the central tendency of the data set. The most common of these measures are the mean, the median, and the mode.

Table 1. Number of days with precipitation in December for Piney and Steinback, 1967-81.

Year	Piney	Steinback
1967	10	12
1968	12	12
1969	9	13
1970	7	15
1971	10	13
1972	11	9
1973	9	16
1974	10	11
1975	9	12
1976	13	13
1977	8	10
1978	9	9
1979	10	13
1980	8	14
1981	9	15
Σ(Xi)	144	187
N	15	15

The mean values of these two sets is determined by summing of the yearly values divided by the number of observations in each data set. In mathematical notation this caiculation would be expressed as:

$$\text{mean } (\bar{x}) = \Sigma(Xi)/N$$

where Xi is the individual values, N is the number of values, and Σ is sigma, a sign used to show summation.

Thus, the calculate means for Piney and Steinback are:

Piney mean = 10 (rounded off)

Steinback mean = 13 (rounded off)

The mode of a data series is that value that occurs with greatest frequency. For Piney, the most frequent value is 9 which occurs five times. The mode for Steinback is 13.

The third measure of central tendency is called the median. The median is the middle value (or the average of the two middle values in an even series) of the data set when the observations are organized in ascending order. For the two locations in question, the medians are:

Piney

9, 9, 10, 11, 12, 12, 12, 13, 13, 13, 13, 14, 15, 15, 16

median = 13

Steinback

7, 8, 8, 9, 9, 9, 9, 9, 10, 10, 10, 10, 11, 12, 13

median = 9

Measures of Dispersion

Measures of central tendency provide no clue into how the observations are dispersed within the data set. Dispersion can be calculated by a variety of descriptive statistics including the range, variance, and standard deviation. The simpest measure of dispersion is the range. The range is calculated by subtracting the smallest individual value from the largest. When presented together with the mean, this statistic provides a measure of data set variability. The range, however, does not provide any understanding to how the data are distributed about the mean. For this measurement, the standard deviation is of value.

The following information describes the calculation of the range, variance, and standard deviation for the data set in Table 2.

Table 2. Dates of the first fall frost at Somewhere, USA, for an 11-year period.

Day of First Frost *	$X_i - \bar{x}$	$(X_i - \bar{x})^2$
291	—8	64
299	0	0
279	—20	400
302	3	9
280	—19	361
303	4	16
299	0	0
304	5	25
307	8	64
314	15	225
313	14	196
$\Sigma(X_i) = 3291$		
$\bar{x} = 3291/11 = 299$		$\Sigma(X_i - \bar{x})^2 = 1360$

*The dates are given in year days, i.e., January 1st is day 1, January 2nd is day 2, and so on throughout the year.

The range for the data is set is derived by subtracting 279 (the smallest value) from 314 (the largest value). The range is 35 days.

The first step in the calculation of standard deviation is to determine the variance by obtaining the deviations of the individual values (Xi) from the mean ($\bar{x}$). The formula for variance (S^2) is:

$$S^2 = [(X_i - \bar{x})^2] / (N-1)$$

where Σ is the summation sign, $(X_i - \bar{x})^2$ is calculated (third column), and N is the number of observations. Standard deviation (S) is merely the square root of the variance ($\sqrt{S^2}$).

In the case of the Somewhere data, the standard deviation is:

$$S^2 = 1356 / 10$$

$$S = 11.6 \text{ or } 12 \text{ (to the nearest day)}$$

This value provides significant information about the distribution of data around the mean. For example:

(a) The mean ± one sample standard deviation contains approximately 68 % of the measurements in the data series.

(b) The mean ± two sample standard deviations contains approximately 95 % of the measurements in the data series.

In Somewhere, the corresponding dates for fall frosts ± one and two standard deviations from the mean (day 299) are:

Minus two standard deviations: 299 - 24 = 275

Minus one standard deviation: 299 - 12 = 287

Plus one standard deviation: 299 + 12 = 311

Plus two standard deviations: 299 + 24 = 323

The calculations above suggest that the chance of frost damage is only 2. 5 % on October 2nd (day 275), 16 % on October 15th (day 287), 50 % on October 27th (day 299), 84 % on November 8th (day 311), and 97. 5 % on November 20th (day 323).

Hypothesis Testing

A hypothesis can be defined as a tentative assumption that is made for the purpose of empirical scientific testing. A hypothesis becomes a theory of science when repeated testing produces the same conclusion. In most cases, hypothesis testing involves the following structured sequence of steps. The first step is the formulation of a null hypothesis. The null hypothesis is the assumption that will be maintained by the researcher unless the analysis of data provides significant evidence to disprove it.

The null hypothesis is denoted symbolically as H0. For example, here is a formulated null hypothesis related to the investigation of precipitation patterns over adjacent rural and urban land-use types:

H0 : There is no difference in precipitation levels between urban and adjacent rural areas.

The second step of hypothesis testing is to state the *alternative hypothesis* (H1). Researchers should structure their tests so that all outcomes are anticipated before the tests and that results can be clearly interpreted. Some tests may require the formulation of multiple alternative hypotheses. However, interpretation is most clear cut when the hypothesis is set up with only one alternative outcome. For the example dealing with precipitation patterns over adjacent rural and urban land-use types, the alternative might be:

H1 : There is an increase in precipitation levels in urban areas relative to adjacent rural areas because of the heating differences of the two surface types (the urban area heats up more and has increased convective uplift).

Step three involves the collection of data for hypothesis testing. It is assumed that this data is gathered in an unbiased manner. For some forms of analysis that use inferential statistical tests the data must be collected randomly, data observations should be independent of each other, and the variables should be normally distributed.

The fourth step involves testing the null hypothesis through predictive analysis or via experiments. The results of the test are then interpreted (acceptance or rejection of the null hypothesis) and a decision may be made about future investigations to better understand the system under study. In the example used here, future investigations may involve trying to determine the mechanism responsible for differences in precipitation between rural and urban land-use types.

Inferential Statistics and Significance Levels

Statisticians have developed a number of mathematical procedures for the purpose of testing hypotheses. This group of techniques is commonly known as *inferential statistics*. Inferential statistics are available both for predictive and experimental hypothesis testing. This group of statistical procedures allow researchers to test assumptions about collected data based on the laws of probability.

Tests are carried out by comparing calculated values of the test statistic to assigned critical values. For a given null hypothesis, the calculated value of the test statistic is compared with tables of critical values at specified significance levels based on probability. For example, if a calculated test statistic exceeds the critical value for a significance level of 0.05 then this means that values of the test statistic as large as, or larger than calculated from the data would occur by chance less than 5 times in 100 if the null hypothesis was indeed correct.

In other words, if we were to reject the null hypothesis based on this probability value of the test statistic, we would run a risk of less than 5 % of acting falsely.

One-tailed and Two-tailed Tests

When using some types of inferential statistics the alternative hypothesis may be directional or non-directional. A directional hypothesis (or one-sided hypothesis) is used when either only positive or negative differences are of interest in an experimental study. For example, when an alternative hypothesis predicts that the mean of one sample would be greater (but not less) than another, then a directional alternative would be used. This type of statistical procedure is known as a one-tailed test. A non-directional (or two-sided) hypothesis would be used when both positive and negative differences are of equal importance in providing evidence with which to test the null hypothesis. We call this type of test two-tailed.

Inferential Statistics: Regression and Correlation

Regression and correlation analysis are statistical techniques used extensively in physical geography to examine causal relationships between variables. Regression and correlation measure the degree of relationship between two or more variables in two different but related ways. In regression analysis, a single dependent variable, Y, is considered to be a function of one or more independent variables, X1, X2, and so on. The values of both the dependent and independent variables are assumed as being ascertained in an error-free random manner.

Further, parametric forms of regression analysis assume that for any given value of the independent variable, values of the dependent variable are normally distributed about some mean. Application of this statistical procedure to dependent and independent variables produces an equation that "best" approximates the functional relationship between the data observations.

Correlation analysis measures the degree of association between two or more variables. Parametric methods of correlation analysis assume that for any pair or set of values taken under a given set of conditions, variation in each of the variables is random and follows a normal distribution pattern. Utilization of correlation analysis on dependent and independent variables produces a statistic called the correlation coefficient (r). The square of this statistical parameter (the coefficient of determination or r^2) describes what proportion of the variation in the dependent variable is associated with the regression of an independent variable.

Analysis of variance is used to test the significance of the variation in the dependent variable that can be attributed to the regression of one or more independent variables. Employment of this statitical procedure produces a calculated F-value that is compared to a critical F-values for a particular level of statistical probability. Obtaining a significant calculated F-value indicates that the results of regression and correlation are indeed true and not the consequence of chance.

Simple Linear Regression

In a simple regression analysis, one dependent variable is examined in relation to only one independent variable. The analysis is designed to derive an equation for the line that best models the relationship between the dependent and independent variables. This equation has the mathematical form:

$$Y = a + bX$$

where, Y is the value of the dependent variable, X is the value of the independent variable, a is the intercept of the regression line on the Y axis when X = 0, and b is the slope of the regression line.

The following table contains randomly collected data on growing season precipitation and cucumber yield (Table 3). It is reasonable to suggest that the amount of water received on a field during the growing season will influence the yield of cucumbers growing on it. We can use this data to illustate how regression analysis is carried out. In this table, precipitation is our independent variable and is not affected by variation in cucumber yield. However, cucumber yield is influenced by precipitation, and is therefore designated as the Y variable in the analysis.

Table 3. Cucumber yield vs precipitation data for 62 observations.

Precipitation mm (X)	Cucumbers kilograms per m^2 (Y)	Precipitation mm (X)	Cucumbers kilograms per m^2 (Y)
22	.36	103	.74
6	.09	43	.64
93	.67	22	.50
62	.44	75	.39
84	.72	29	.30
14	.24	76	.61
69	.61	29	.38
104	.66	50	.53
100	.80	59	.58
41	.47	70	.62

85	.60	81	.66
90	.51	93	.69
27	.14	99	.71
18	.32	14	.14
48	.21	51	.41
37	.54	75	.66
67	.70	6	.18
56	.67	20	.21
31	.42	36	.29
17	.39	50	.56
7	.25	9	.13
2	.06	2	.10
53	.47	21	.18
70	.55	17	.17
6	.07	87	.63
90	.69	97	.66
46	.42	33	.18
36	.39	20	.06
14	.09	96	.58
60	.54	61	.42

$\Sigma X = 3{,}050$

$\bar{x} = 49.1935$

$\Sigma Y = 26.62$

$\bar{y} = 0.4294$

$n = 62$

Often the first step in regression analysis is to plot the X and Y data on a graph (Figure 3). This is done to graphically visualize the relationship between the two variables. If there is a simple relationship, the plotted points will have a tendancy to form a recognizable pattern (a straight line or curve). If the relationship is strong, the pattern will be very obvious. If the relationship is weak, the points will be more spread out and the pattern less distinct. If the points appear to fall pretty much at random, there may be no relationship between the two variables.

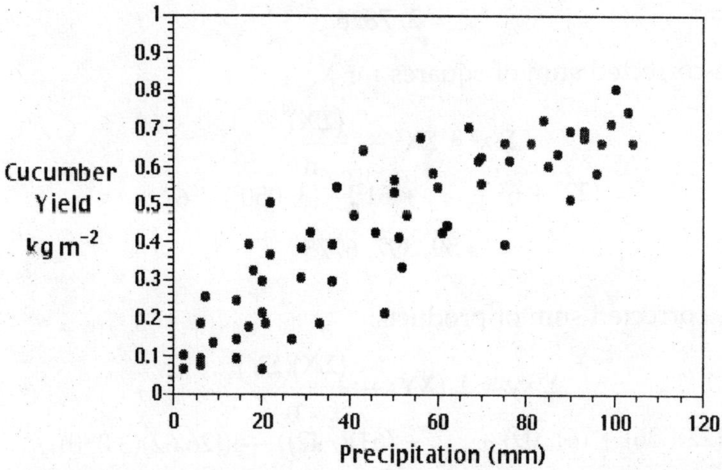

Figure 3. Scattergram plot of the precipitation and cucumber yield data found in Table 3. The distribution of the data points indicates a possible positive linear relationship between the two variables.

The type of pattern (straight line, parabolic curve, exponential curve, etc.) will determine the type of regression model to be applied to the data. In this particular case, we will examine data that produces a simple straight-line relationship (Figure 3). After selecting the model to be used, the next step is to calculate the corrected sums of squares and products used in a bivariate linear regression analysis. In the following equations, capital letters indicate uncorrected values of the variables and lower-case letters are used for the corrected parameters in the analysis.

The corrected sum of squares for Y:

$$\Sigma y^2 = \Sigma Y^2 — \frac{(\Sigma Y)^2}{n}$$

$$= (0.36^2 + 0.09^2 + \ldots + 0.42^2) — (26.622) / 62$$

$$= 2.7826$$

The corrected sum of squares for X:

$$\Sigma x^2 = \Sigma X^2 — \frac{(\Sigma X)^2}{n}$$

$$= (22^2 + 6^2 + \ldots + 61^2) - (3,050^2) / 62$$

$$= 59,397.6775$$

The corrected sum of products:

$$\Sigma xy = \Sigma (XY) — \frac{(\Sigma X)(\Sigma Y)}{n}$$

$$= ((22)(.36) + (6)(.09) + \ldots + (61)(.42)) — ((26.62)(3,050))/62$$

$$= 354.1477$$

As discussed earlier, the general form of the equation for a straight line is Y = a + bX. In this equation, a and b are constants or regression coefficients that are estimated from the data set. Based on the mathematical procedure of least squares, the best estimates of these coefficients are:

$$b = \frac{(\Sigma xy)}{(\Sigma x^2)}$$

$$= (354.1477) / (59,397.6775) = 0.0060$$

$$a = Y — bX = 0.42935 — (0.0060)(49.1935) = 0.1361$$

Substituting these estimates into the general linear equation suggests the following relationship between the Y and X variables:

$$\hat{Y} - 0.1361 + 0.0060X$$

where $\hat{Y}$ indicates that we are using an estimated value of Y.

With this equation, we can estimate the the number of cucumbers (Y) from the measurements of precipitation (X) and describe this relationship on our scattergram with a best fit straight-line (Figure 4). Because Y is estimated from a known value of X, it is called the dependent variable and X the independent variable. In plotting the data in a graph, the values of Y are normally plotted along the vertical axis and the values of X along the horizontal axis.

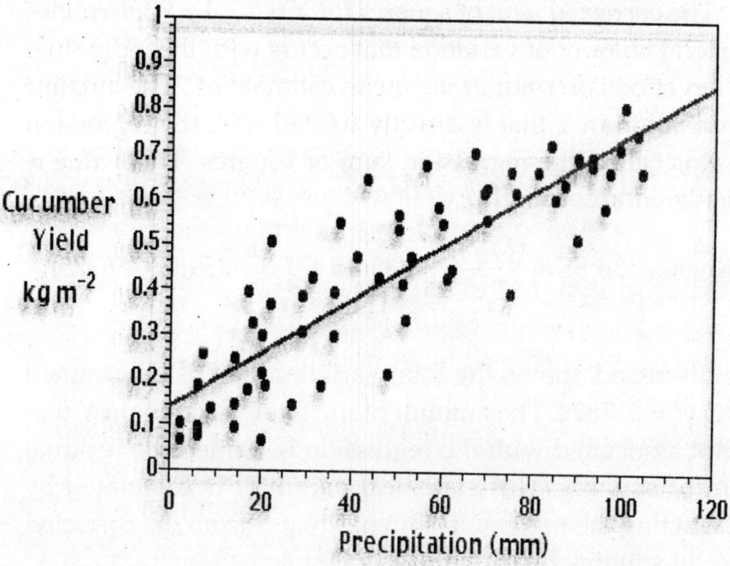

Figure 4: Scattergram plot of the precipitation and cucumber yield data and the regression model best fit straight-line describing the linear relationship between the two variables.

Regression Analysis and ANOVA

A regression model can be viewed of as a type of moving average. The regression equation attempts to explain the relationship between the Y and X variables through linear association. For a particular value of X, the regression model provides us with an estimated value of Y. Yet, Figure 4 indicates that many of the plotted values of the actual data are observed to be above the regression line while other values are found below it. These variations are caused either by sampling error or the fact that some other unexplained independent variable influences the individual values of the Y variable.

The corrected sum of squares for Y (i. e. , Σy^2) determines the total amount of variation that occurs with the individual observations of Y about the mean estimate of . The amount of variation in Y that is directly related with the regression on X is called the regression sum of squares. This value is calculated accordingly:

$$\text{Regression SS} = \frac{(\Sigma xy)^2}{\Sigma x^2} = (354.\ 1477)2\ /\ (59,\ 397.\ 6775)$$

$$= 2.\ 1115$$

As discussed above, the total variation in Y is determined by $\Sigma\, y^2 = 2.\ 7826$. The amount of the total variation in Y that is not associated with the regression is termed the residual sum of squares. This statistical paramter is calculated by subtracting the regression sum of squares from the corrected sum of squares for Y ($\Sigma\, y^2$):

$$\text{Residual SS} = \Sigma\, y^2 — \text{Regression SS}$$

$$= 2.\ 7826 - 2.\ 1115$$

$$= 0.\ 6711$$

The unexplained variation can now be used as a standard for testing the amount of variation attributable to the regression. Its significance can be tested with the F test from calculations performed in an Analysis of Variance table.

Source of variation	df 1	SS	MS 2
Due to regression	1	2. 1115	2. 1115
Residual (unexplained)	60	0. 6711	0. 0112
Total	61	2. 7826	-

1 There were 62 values of Y analyzed and therefore n = 62. The total sum of squares degrees of freedom (df) is determined as n-1 or 61. The regression of Y on X has 1 degree of freedom. The residual or unexplained degrees of freedom is determined by subtracting regression df (1) from total sum of squares df (61).

2 MS is calculated as SS / df.

Using the Analysis of Variance procedure, the regression is tested by determining the calculated F statistic:

$$F = (\text{Regression MS}) / (\text{Residual SS}) = (2.\ 1115) / (0.\ 0112) = 188.\ 86$$

To test this statistic we use a table of F to determine a critical test value for a probability of 0. 01 or 1 % (this relationship can occur by chance only in 1 out 100 cases) and with 1, 60 degrees of freedom. According to the table the critical test value is 7.1. In this test, the relationship is deemed significant if the calculated F statistic is greater than the critical test value. This regression is statistically significant at the 0. 01 level because 188.86 is greater than 7.1.

Caution must be taken when interpreting the results of regression. In our example, we found a significant relationship between precipitation and cucumber yield. However, this conclusion may not be the result of a causal relationship between the two variables. A third variable that is directly associated to both precipitation and cucumber yield may be confounding the interpretation of the analysis. Absolute verification of associations between variables can only be confirmed with experimental manipulation.

Coefficient of Determination

To measure how strong the correlation is between the two variables, we can determine the amount of the total variation in Y that is associated with the regression model. This ratio is sometimes called the coefficient of determination and is represented by the symbol r^2. The value of the coefficient of determination ranges from 1.00 to 0.00. The calculated coefficient of determination from the data set above was 0.76 or 76 % (as calculated below). This value suggests that 76 % of the variation in Y was associated with the change seen X from the data set observations.

$$\text{Coefficient of determination} = (\text{Regression SS}) / (\text{Total SS})$$

$$= (2.1115) / (2.7826) = 0.7588$$

Correlation Coefficient

Another useful regression statistic that measures the strength of the correlation between to variables is the correlation coefficient. This statistic is often represented by the symbol "r" and is determined by taking the square-root of the coefficient of determination. The value of the correlation coefficient ranges from 1.00 to -1.00. A value of 0. 0 indicates that there is absolutely no relationship between the X and Y variables. The strength of the relationship between the X and Y variables increases as the value of r approaches 1.00 and -1.00. Perfect correlation occurs if r equals either 1.00 (perfect positive) or -1.00 (perfect negative). Positive correlation coefficients indicate that an increase in the value of the X variable results in an increase in the value of the Y variable. Negative correlation coefficients indicate that an increase in the value of the X variable results in a decrease in the value of the Y variable.

2

Theories of Physical Geography

The world of nature is very complex. In order to understand this complexity humans usually try to visualize the phenomena of nature as a system. A system is a set of interrelated components working together towards some kind of process. One of the simplest forms of a system is a model. Both models and systems are simplified versions of reality. In Physical Geography, and many other fields of knowledge, systems and models are used extensively as aids in explaining natural phenomena around us.

Asystem is a group of parts that interact according to some kind of process. Systems are often visualized or modeled as component blocks with some kind of connections drawn.

All systems have the same common characteristics. These common characteristics are summarized below:

— All systems have some structure.

— All systems are generalizations of reality.

— They all function in the same way.

— There are functional as well as structural relationships between the units of a system.

— Function implies the flow and transfer of some material. Systems exchange energy and matter internally and with their surrounding environment through various processes of input and output.

— Function requires the presence of some driving force, or some source of energy.

— All systems show some degree of integration.

Within its defined boundary the system has three kinds of properties:

— *Elements* : are the kinds of things or substances composing the system. They may be atoms or molecules, or larger bodies of matter-sand grains, rain drops, plants, or cows.

— *Attributes* : are characteristics of the elements that may be perceived, for example: quantity, size, color, volume, temperature, and mass.

— *Relationships*: are the associations that exist between elements and attributes based on cause and effect.

The state of the system is defined when each of its properties (e.g. elements, attributes, and relationships) has a defined value. Scientists have examined and classified many types of systems. These types include:

— *Isolated System*: a system where there are no interactions outside its boundary layer. Such systems are common in laboratory experiments.

— *Closed System*: is closed with respect to matter, but energy may be transferred between the system and its surroundings. Earth is essentially a closed system.

— *Open System* : is a system where both matter and energy can cross the boundary of the system. Most environmental systems are open.

— *Morphological System*: is a system where we understand process relationships or correlations between the elements of the system in terms of measured features.

— *Cascading System*: in these systems we are interested in the movement of energy and/or matter from one element to another and understand the processes involved.

— *Process-Response System*: in these systems we are interested in the movement, storage and transformation of energy and matter and the relationships between measured features in the various elements of the system.

— *Control System*: a system that is intelligently manipulated by humans.

— *Ecosystem*: is concerned with the biological relationships within the environment and the interactions between organisms and their physical surroundings.

Systems exist at every scale of size and are often arranged in some kind of hierarchical fashion. Large systems are often composed of one or more smaller systems working within its various elements. Processes within these smaller systems can often be connected directly or indirectly to processes found in the larger system. A good example of a system within systems is the hierarchy of systems found in our Universe.

At the highest level in this hierarchy we have the system that we call the cosmos or Universe. The elements of this system consist of galaxies, quasars, black holes, stars, planets and other heavenly bodies. The current structure of this system is thought to have come about because of a massive explosion known as the Big Bang and is controlled by gravity, weak and strong atomic forces, and electromagnetic forces. Around some stars in the universe we have an obvious arrangement of planets, asteroids, comets and other material. We call these systems solar systems.

The elements of this system behave according to set laws of nature and are often found orbiting around a central star because of gravitational attraction. On some planets conditions may exist for the development of dynamic interactions between the hydrosphere, lithosphere, atmosphere, or biosphere. We can define a planetary system as a celestial body in space that orbits a star and that maintains some level of dynamics between its lithosphere, atmosphere and hydrosphere. Some planetary systems, like the Earth, can also have a biosphere.

If a planetary system contains a biosphere, dynamic interactions will develop between this system and the lithosphere, atmosphere and hydrosphere. These interactions can be called an environmental system. Environmental systems can also exist at smaller scales of size (e.g., a single flower growing in a field could be an example of a small-scale environmental system). The Earth's biosphere is made up small interacting entities called ecosystems. In an ecosystem, populations of species group together into communities and interact with each other and the abiotic environment. The smallest lining entity in an ecosystem is a single organism.

An organism is alive and functioning because it is a biological system. The elements of a biological system consist of cells and larger structures known as organs that work together to produce life. The functioning of cells in any biological system is dependent on numerous chemical reactions. Together these chemical reactions make up a chemical system. The types of chemical interactions found in chemical systems are dependent on the atomic structure of the reacting matter. The components of atomic structure can be described as an atomic system.

Environmental systems also involve the capture, movement, storage, and use of energy. Thus, environmental systems are also energy systems. In environmental systems, energy moves from the abiotic environment to life through processes like plant photosynthesis. Photosynthesis packages this energy into simple organic compounds like glucose and starch. Both of these organic molecules can be stored for future use. The following chemical formula describes how plants capture the sun's light energy and convert it into chemical energy:

$$6CO_2 + 6H_2O + \text{light energy} = C_6H_{12}O_6 + 6O_2$$

The energy of light is used by plants in this reaction to chemically change carbon dioxide (CO_2) and water (H_2O) into oxygen (O_2) and the energy rich organic molecule glucose ($C_6H_{12}O_6$). The chemical energy of photosynthesis can be passed on to other living or biotic components of an environmental system through biomass consumption or decomposition by consumer organisms.

When needed for metabolic processes, the fixed organic energy stored in an organism can be released to do work via respiration or fermentation. Energy also fuels a number of environmental processes that are essentially abiotic. For example, the movement of air by wind, the weathering of rock into soil, the formation of precipitation, and the creation of mountains by tectonic forces. The first three processes derive their energy directly or indirectly from the sun's radiation that is received at the Earth's surface. Mountain building is fueled by the heat energy that exists within the Earth's interior. Finally, the movement of energy in environmental systems always obeys specific thermodynamic laws that cannot be broken.

Concepts of Equilibrium

Equilibrium describes the average condition of a system, as measured through one of its elements or attributes, over a specific period of time. There are six types of equilibrium:

1. Steady state equilibrium is an average condition of a system where the trajectory remains unchanged in time.

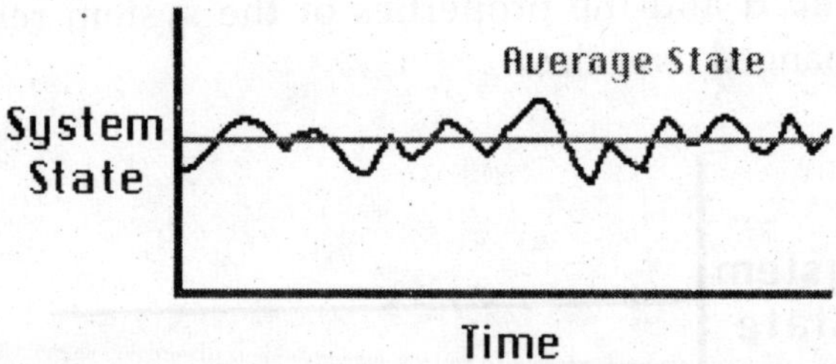

Figure 1: Example of the state of a steady state equilibrium over time.

2. Thermodynamic equilibrium describes a condition in a system where the distribution of mass and energy moves towards maximum entropy.

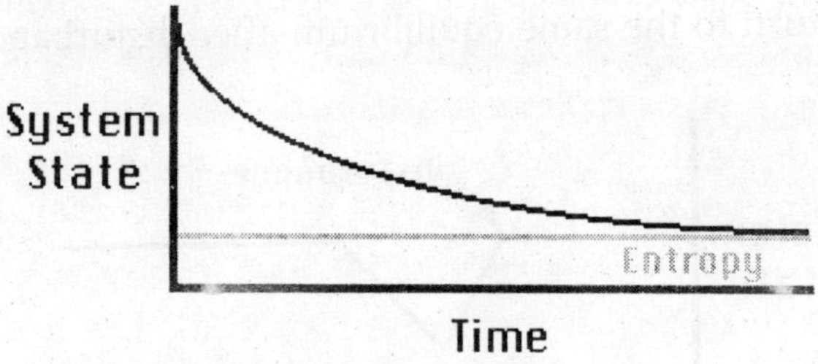

Figure 2: Example of the state of a thermodynamic equilibrium over time.

3. A dynamic equilibrium occurs when there are unrepeated average states through time.

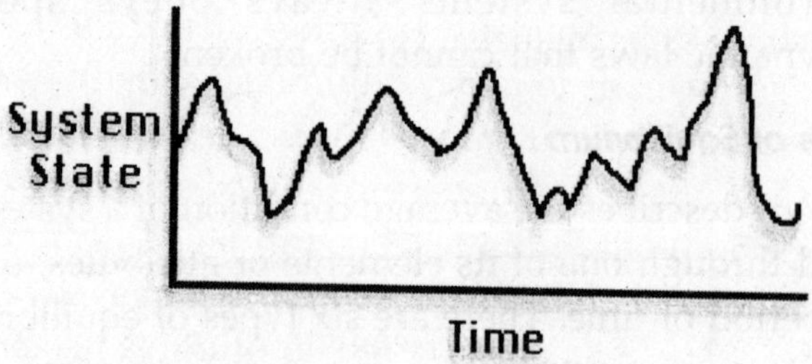

Figure 3: Example of the state of a dynamic equilibrium over time.

4. Static equilibrium occurs where force and reaction are balanced and the properties of the system remain unchanged over time.

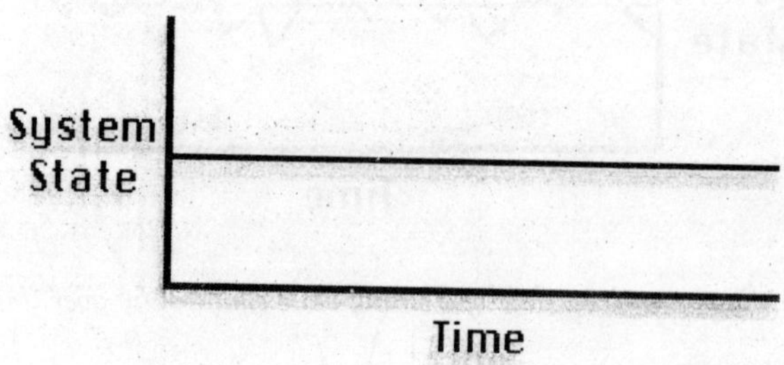

Figure 4: Example of the state of a static equilibrium over time.

5. In a stable equilibrium the system displays tendencies to return to the same equilibrium after disturbance.

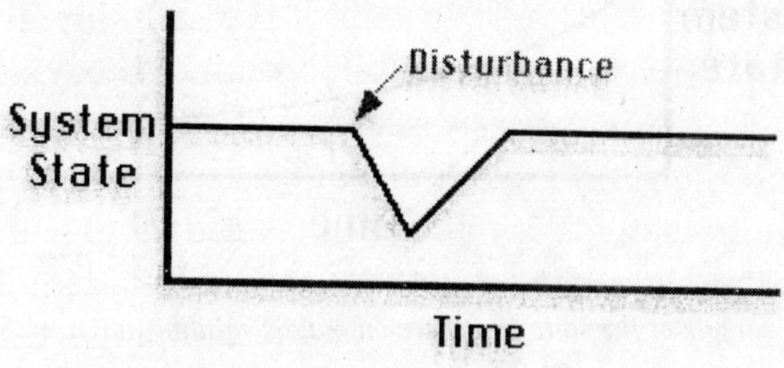

Figure 5: Example of the state of a stable equilibrium over time.

6. In an unstable equilibrium the system returns to a new equilibrium after disturbance.

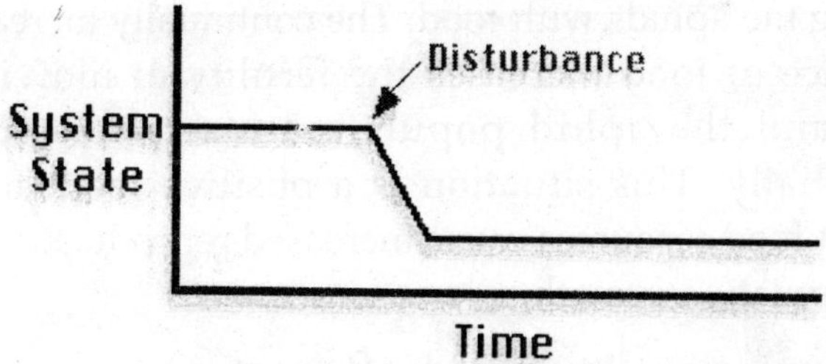

Figure 6: Example of the state of an unstable equilibrium over time.

In order for a system to maintain a steady state or average condition the system must possess the capacity for self-regulation. Self-regulation in many systems is controlled by negative feedback and positive feedback mechanisms. Negative-feedback mechanisms control the state of the system by dampening or reducing the size of the system's elements or attributes. Positive-feedback mechanisms feed or increase the size of one or more of the system's elements or attributes over time.

Interactions among living organisms, or between organisms and the abiotic environment typically involve both positive feedback and negative feedback responses. Feedback occurs when an organism's system state depends not only on some original stimulus but also on the results of its previous system state. Feedback can also involve the system state of non-living components in an ecosystem. A positive feedback causes a self-sustained change that increases the state of a system. Negative feedback causes the system to decease its state over time. The presence of both negative and positive feedback mechanisms in a system results in self-regulation.

Over the winter only a few aphids survive the cold winter temperatures and scarcity of food. However, warmer temperatures in the spring cause plants to start growing, providing the aphids with food. The continually increasing abundance of food increases the fertility of individual aphids and the aphid population starts to expand exponentially. This situation is a positive feedback as abundant food resources cause increased reproduction and rapid population growth.

By late summer the supply of plants has reached its maximum. As a result, the availability of food to the growing aphid population becomes less per individual as time proceeds. Less food means lower egg production and births begin to decline. When births reach the same level as deaths the aphid population stops growing and stabilizes. The population is becoming too large for the food supply. The size of the population relative to food supply produces a negative feedback as the fertility of aphids begins to decline.

With the arrival of fall cooler temperatures reduce the supply of plants causing the large aphid population to be faced with further food scarcity. The lack of food causes the aphid death rate to increase and the birth rate to decrease. The population begins to decline rapidly. The reduction of food caused by the cooler temperatures enhances the negative feedback that began in late summer.

About 11 to 15 billion years ago all of the matter and energy in the Universe was concentrated into an area the size of an atom. At this instant, matter, energy, space and time did not exist. Then suddenly, the Universe began to expand at an incredible rate and matter, energy, space and time came into being (the Big Bang). As the Universe expanded, matter began to coalesce into gas clouds, and then stars and planets. Our solar system formed about 5 billion years ago when the Universe was about 65 % of its present size. Today, the Universe continues to expand.

Big Bang Theory

The acceptance of this theory by the scientific community is based on a number of observations. These observations confirm specific predictions of the Big Bang theory. Predictions associated with the Big Bang theory that have been tested by this process are: If the Big Bang did occur, all of the objects within the Universe should be moving away from each other. In 1929, Edwin Hubble documented that the galaxies in our Universe are indeed moving away from each other.

The Big Bang should have left an "afterglow" from the explosion. In the 1960s, scientists discovered the existence of cosmic background radiation; the so-called "afterglow" after the Big Bang explosion. Our most accurate measurements of this cosmic radiation came in November 1989, by the Cosmic Background Explorer (COBE) satellite. The measurements from this satellite tested an important prediction of the Big Bang theory. This prediction suggests that the initial explosion that gave birth to the Universe should have created radiation with a spectrum that follows a blackbody curve.

The COBE measurements indicated that the spectrum of the cosmic radiation varied from a blackbody curve by only 1 %. This level of error is considered insignificant. If the Universe began with a Big Bang, extreme temperatures should have caused 25 percent of the mass of the Universe to become helium. This is exactly what is observed. Matter in the Universe should be distributed homogeneously. Astronomical observations from the Hubble Space Telescope do indicate that matter in the Universe generally has a homogeneous distribution.

Cosmologists have postulated two endings to the Universe. If the Universe is infinite or has no edge, it should continue to expand forever. A Universe that is finite or closed is theorized to collapse when expansion stops because of gravity. The collapse of the Universe ends when all matter and energy is compressed into the high energy, high-density state from which it began. This scenario is of course called the Big Crunch. Some theorists have suggested that the Big Crunch will produce a new Big Bang and the process of an expanding Universe will begin again. This idea is called the oscillating Universe theory.

History of the Earth

Scientists believe the Earth began its life about 4.6 billion years ago. The Earth formed as cosmic dust lumped together to form larger and larger particles until 150 million years had passed. At about 4.4 billion years, the young Earth had a mass similar to the mass it has today. The continents probably began forming about 4.2 billion years ago as the Earth continued to cool.

The cooling also resulted in the release of gases from the lithosphere, much of which formed the Earth's early atmosphere. Most of the Earth's early atmosphere was created in the first one million years after solidification (4.4 billion years ago). Carbon dioxide, nitrogen, and water vapor dominated this early atmosphere.

As the Earth continued to cool, the water vapor found in the atmosphere condensed to form the oceans and other fresh water bodies on the continents. Oxygen began accumulating in the atmosphere through photo-dissociation of 02 from water, and by way of photosynthesis (life). The emergence of living organisms was extremely important in the creation of atmospheric oxygen and ozone. Without ozone, life could not exist on land because of harmful ultraviolet radiation.

Most of the build up of oxygen in the atmosphere occurred between 2.1 and 1.5 billion years ago as a direct result of photosynthesis from ocean based plants like algae. At about 450 million years ago, there was enough oxygen in the atmosphere to allow for the development of a stratospheric ozone layer that was thick enough to keep terrestrial life protected from ultraviolet radiation. As a result, terrestrial life began its development and expansion at this time.

Natural Spheres

From the standpoint of Physical Geography, the Earth can be seen to be composed of four principal components:

1. *Lithosphere*: describes the solid inorganic portion of the Earth (composed of rocks, minerals and elements). It can be regarded as the outer surface and interior of the solid Earth. On the surface of the Earth, the lithosphere is composed of three main types of rocks:

   — Igneous - rocks formed by solidification of molten magma.

   — Sedimentary - rocks formed by the alteration and compression of old rock debris or organic sediments.

— Metamorphic - rocks formed by alteration of existing rocks by intense heat or pressure.

2. *Atmosphere* : is the vast gaseous envelope of air that surrounds the Earth. Its boundaries are not easily defined. The atmosphere contains a complex system of gases and suspended particles that behave in many ways like fluids. Many of its constituents are derived from the Earth by way of chemical and biochemical reactions.

3. *Hydrosphere* : describes the waters of the Earth (see the hydrologic cycle). Water exists on the Earth in various stores, including the atmosphere, oceans, lakes, rivers, soils, glaciers, and groundwater. Water moves from one store to another by way of: evaporation, condensation, runoff, precipitation, infiltration and groundwater flow.

4. *Biosphere*: consists of all living things, plant and animal. This zone is characterized by life in profusion, diversity, and ingenious complexity. Cycling of matter in this sphere involves not only metabolic reactions in organisms, but also many abiotic chemical reactions.

All of these spheres are interrelated to each other by dynamic interactions, like biogeochemical cycling, that move and exchange both matter and energy between the four components.

Gaia Hypothesis

In 1965, J.E. Lovelock published the first scientific paper suggesting the Gaia hypothesis. The Gaia hypothesis states that the temperature and composition of the Earth's surface are actively controlled by life on the planet. It suggests that if changes in the gas composition, temperature or oxidation state of the Earth are caused by·extraterrestial, biological, geological, or other disturbances, life responds to these changes by modifying the abiotic environment through growth and metabolism. In simplier terms, biological responses tend to regulate the state of the Earth's environment in their favor.

This theory is important to Physical Geography and other Earth Sciences for the following reasons:

— The Gaia theory suggests that the abiotic and biotic environment is made up of many complex interrelationships;

— Many of these complex interrelationships are quite delicate and may be altered by human activity to a breaking point; and

— The theory suggests that humans must learn to respect Gaia by reducing their intentional modification of the Earth's abiotic and biotic components.

3

Understanding Energy and Matter

Energy is defined simply by scientists as the capacity for doing work. Matter is the material (atoms and molecules) that constructs things on the Earth and in the Universe. Albert Einstein suggested early in this century that energy and matter are related to each other at the atomic level. Einstein theorized that it should be possible to convert matter into energy. From Einstein's theories, scientists were able to harness the energy of matter beginning in the 1940s through nuclear fission. The most spectacular example of this process is a nuclear explosion from an atomic bomb.

A more peaceful example of our use of this fact of nature is the production of electricity from controlled fission reactions in nuclear reactors. Einstein also suggested that it should be possible to transform energy into matter. Energy and matter are also associated to each other at much larger scales of nature.

Types of Energy

Energy comes in a variety forms. The simplest definition of the types of energy suggests that two forms exist: kinetic energy and potential energy. Kinetic energy is the energy due to motion. A rock falling from a cliff, a bee in flight, wind blowing leaves of trees, and water following over a waterfall are all examples of kinetic energy. Potential energy is the energy stored by an object that can be potentially transformed into another form of energy. Water stored behind a dam, the chemical energy of the food we consume, and the gasoline that we putting in our cars are all examples of potential energy.

Conversion of this energy occurs when the energy in food is used for metabolism and it has been digested, when the water in the dam flows through turbines to produce electricity from motion, and when the gasoline is used in a engine to produce motion from combustion. Some other forms of energy include heat, electricity, sound, energy of chemical reactions, magnetic attraction, energy of atomic reactions, and light. Definitions for a few of these types of energy are as follows:

— *Radiation*: is the emission and propagation of energy in the form of electromagnetic waves.

— *Chemical Energy*: is energy used or released in chemical reactions.

— *Atomic Energy*: is energy released from an atomic nucleus at the expense of its mass.

— *Electromagnetic Energy*: is energy stored in electromagnetic waves or radiation. Energy is released when the waves are absorbed by a surface. Any object with a temperature above absolute zero (-273° Celsius) emits this type of energy. The intensity of energy released is a function of the temperature of the radiating surface. The higher the temperature, the greater the quantity of energy released.

— *Electrical Energy*: is energy resulting from the force between two objects having the physical property of charge.

— *Heat Energy*: a form of energy representing aggregated internal energy of motions of atoms and molecules in a body.

On Earth, there are fundamentally three ways in which energy can be transferred from one place to another: conduction, convection, and radiation. Conduction consists of energy transferred directly from molecule to molecule, and represents the flow of energy along a temperature gradient. Convection involves the transfer of energy by means of vertical mass motions of the medium through which heat is transferred (horizontal transfer is called advection).

In gases and liquids, this exchange by mass motions is commonly seen in rising bubbles known as convection currents. Both conduction and convection depend on a material medium in order to operate. This medium can be gaseous, liquid, or solid. Radiation is the only means of energy transfer through space without the aid of a material medium, and is the major source of energy on the Earth.

Matter

Matter is the material that makes up things in the Universe. All matter on the Earth is constructed of elements (see WebElements for the periodic table of elements). Chemists have described approximately 115 different elements. Each of these elements have distinct chemical characteristics. The smallest particle that exhibits the unique chemical characteristics of an element is known as an atom.

Atoms are composed of yet smaller particles known as protons, neutrons, and electrons. A proton is a subatomic particle that has significant mass and contributes a single positive electrical charge to an atom. Neutrons also have significant mass but no electrical charge. Electrons are extremely light subatomic particles having a mass that is 1/1840 of a proton. Each electron also has a negative electrical charge. Protons and neutrons make up the nucleus of an atom. As a result, most of an atom's mass is concentrated in the nucleus. Because protons are positively charged the nucleus has a positive charge equal to the number of these subatomic particles.

Electrons are found orbiting outside the nucleus at various distances based on their energy level. The area occupied by the electrons has a negative charge equal to the number of these subatomic particles. If an atom has an equal number of electrons and protons its net electrical charge is zero. If there are more electrons than protons the charge of the atom is negative. Likewise, if there are less electrons than protons the charge of the atom is positive. In both cases, the exact charge is determined by subtracting protons from electrons. As a result, 4 protons minus 6 electrons give an atomic charge of -2. The number of protons found in the nuclei of the different types of elements is unique and is referred to as the atomic number.

All atoms of a specific element have the same number of protons in their nuclei. Atomic mass number is an atom's total number of neutrons and protons. Many elements have unequal numbers of neutrons and protons in their nucleus. An element's atomic weight refers to the total weight of neutrons, protons, and electrons. For example, the atomic weight of aluminum is 26.98. Atomic number describes the number of protons found in an atom. For example, silver has an atomic number of 47 or 47 protons in its atom.

Some elements can have variants containing different numbers of neutrons but similar numbers of protons. We call these variants isotopes. Carbon has two isotopes. Its most common form is carbon-12 which has 6 protons plus 6 neutrons. About 99 % of the carbon on our planet is of this type. The isotope carbon-13 has 6 protons plus 7 neutrons. Carbon-14 is the rarest isotope of carbon containing 8 neutrons. Some isotopes are unstable and their nucleus tends to lose subatomic particles forming an element with a lower atomic mass. This process is known as radioactive decay.

Elements can be classified as being either metals, nonmetals, or metalloids. Metals are elements that usually conduct heat and electricity and are shiny. Nonmetals do not conduct electricity that well and are normally not shiny. Metalloids have characteristics that are in between metals and nonmetals. Elements with a net positive or negative charge are called ions. Chemists indicate the number of positive or negative charges on an ion using a superscript after the element's symbol. For example, calcium has two positive charges and is written as Ca^{2+}. Some common negatively charged ions include nitrate (NO^{3-}), sulfate (SO_4^{2-}), and phosphate (PO_4^{3-}).

Positive and negative ions are electrically attracted to each other. This mutual attraction allows for the bonding of atoms to occur forming structures of matter that are larger than just one atom. When similar atoms bond together they construct molecules. Atoms of different elements joined together form compounds (Figure 1). Sodium chloride (or table salt), is an ionic compound consisting of sodium (Na^+) and chloride (Cl^-). In nature, it forms as a three-dimensional array of oppositely charged ions (Figure 2). Many of the Earth's substances have a molecular structure similar to sodium chloride.

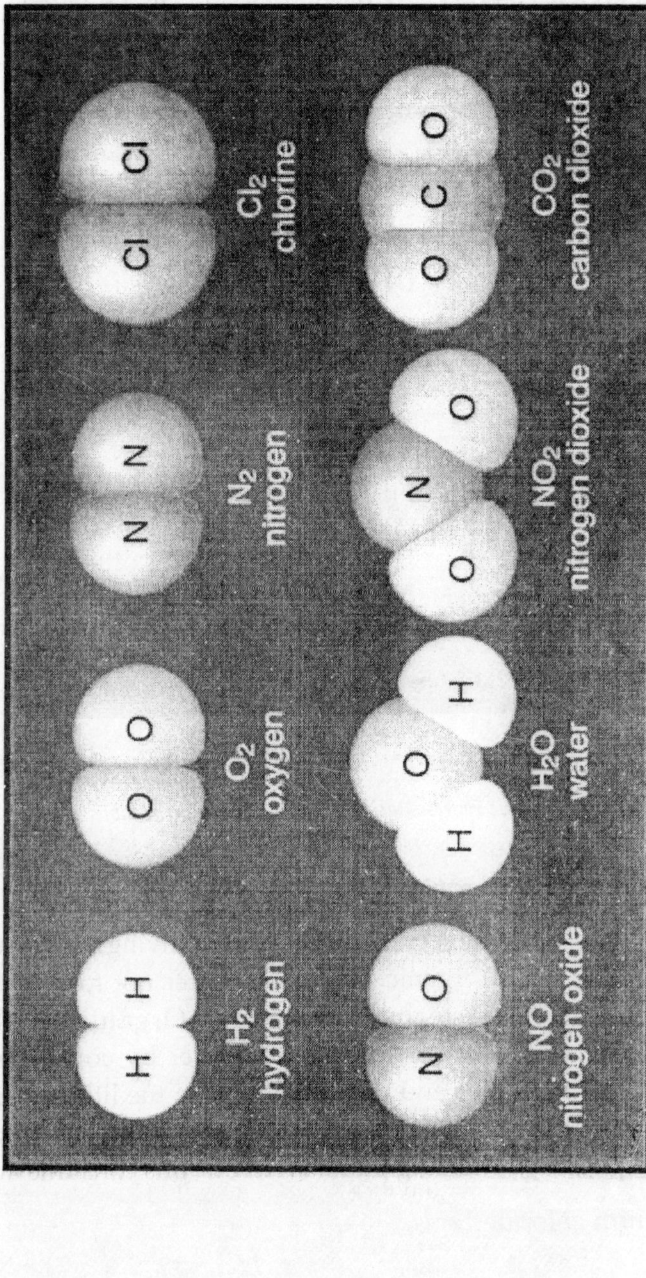

Figure 1: Some common molecules and compounds.

The molecules in the top row bond with each other by sharing electrons. The compounds in the bottom row also share electrons. However, these joins are called ionic bonds.

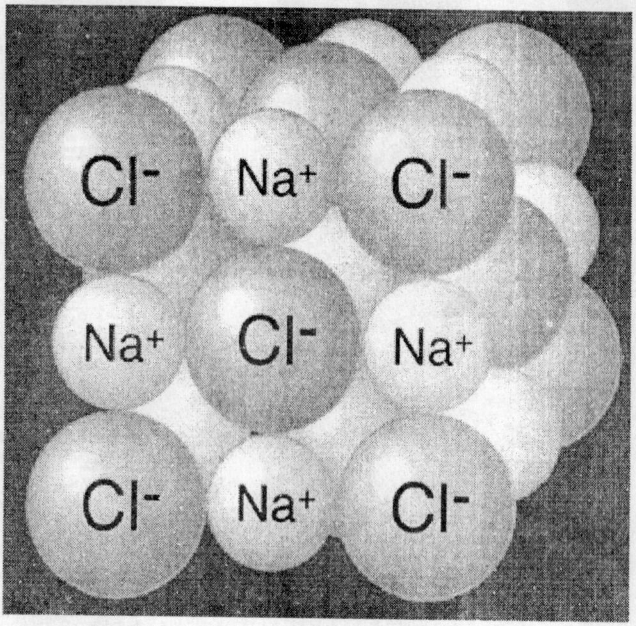

Figure 2: Atomic representation of sodium chloride or table salt.

Inorganic and Organic

Compounds and molecules constructed in living tissues are commonly called organic. Forms of matter not formed by living things are termed inorganic. Organisms like autotrophs usually create organic matter by consuming inorganic molecules and compounds from the lithosphere, hydrosphere, and atmosphere. An example of an autotroph is any photosynthesizing plant. Heterotrophs consume and assimilate other living things to create their organic matter. Herbivores and carnivores are examples of heterotrophs.

There are four general categories of organic compounds: lipids, carbohydrates, proteins, and nucleic acids.

1. *Lipids*: are composed of carbon atoms that have two hydrogen atoms attached. Lipids are commonly known as fats and oils, and belong to the family of molecules known as hydrocarbons.

2. *Carbohydrates*: are composed of carbon, oxygen, and hydrogen atoms. Some examples are sugars, starch, and cellulose.

3. *Proteins*: are organic compounds that are made primarily of carbon, hydrogen, nitrogen, and some other minor elements that are arranged into 20 different compounds known as amino acids.

4. *Nucleic Acids* : are composed primarily of different combinations of carbon, hydrogen, nitrogen, oxygen, and phosphorus. They are very complex compounds being created by the atomic linking of thousands of individual atoms. DNA or deoxyribonucleic acid, the genetic blueprint of life, is an example of a nucleic acid.

Cells

All organisms are composed of one or more of cells. Cells are the smallest self-functioning unit found in living organisms. Cells are also where the processes of metabolism and heredity occur in an organism. Cells arise by the cellular division of a previously existing cell. Biologists have differentiated two basic types of cells in organisms. Bacteria, archaea, and cyanobacteria have cells that are quite uncomplicated in terms of structure and function. Quite

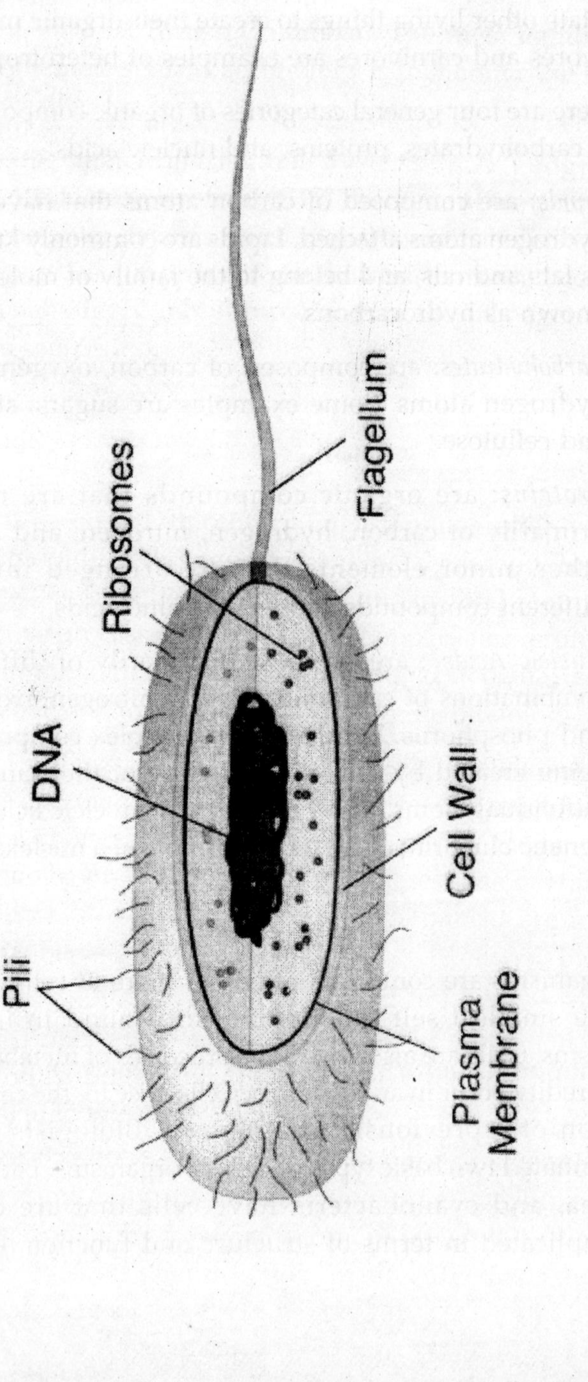

Figure 3: Typical prokaryote cell.

simply, they lack internal organization. These cells are commonly known as prokaryotes (Figure 3).

The cells of plants and animals are more complex than those of bacteria, archaea, and cyanobacteria. We call these type of cells eukaryotes. Eukaryotic cells have a membrane enclosed nucleous which contains the organism's DNA. Plant and animal cells also contain a variety of membrane-bound structures known as organelles.

Eukaryotic cells also show great variation in terms of size. White blood cells of mammals are some of the smallest eukaryotic cells with a diameter between 3 to 4 micrometers. Ostrich ovum are very large cells with a diameter of about 100 micrometers. Internally, cells contain specialized structures known as organelles that carry out distinct cellular functions. Within these structures enzymes, a type of protein, are used to facilitate and regulate various chemical reactions.

Structure of a Cell

— *Cell Wall*: Outer layer on a cell composed of cellulose or other complex carbohydrates. Helps to support and protect the cell.

— *Plasma Membrane*: A layer composed of lipids and proteins that controls the permeability of the cell to water and dissolved substances. Regulates the movement of material into and out of the cell.

— *Flagellum (Flagella pl.)*: Threadlike organelle that extends from the surface of the cell. Found in both prokaryotes and eukaryotes. Used for movement of the cell or to move fluids over the cell's surface for absorption.

— *Pilus (Pili pl.)*: Hollow, hairlike structures made of protein found on prokaryote cells. Allows cell to attach itself to another cell.

— *Ribosomes*: Tiny, complex structures composed of protein and RNA. Often attached to endoplasmic reticulum. Ribosomes are involved in protein synthesis.

— *Endoplasmic*: Reticulum Extensive system of internal membranes. Forms compartments to isolate cell substances.

— *Nucleus*: Double membrane structure that encases chromosomes. Control center of the cell which directs protein synthesis and cell reproduction.

— *Chromosomes*: Long strands of DNA. Store hereditary information.

— *Nucleolus*: Aggregations of rRNA and ribosomal proteins. Area were ribosomes are manufactured.

— *Golgi Complex*: Flattened stacks of membranes. Used in the collection, packaging, and distribution of synthesized molecules.

— *Peroxisomes*: Membrane confined spherical body about 0.2 to 0.5 micrometers in diameter. Formed by the endoplasmic reticulum. Converts fats into carbohydrates. Detoxifies potentially harmful oxidants.

— *Lysosomes*: Membrane confined spherical body about 0.2 to 0.5 micrometers in diameter. Formed by the golgi complex. Contains digestive enzymes for braking down old cellular components.

— *Centrioles*: Long hollow tubes composed of protein. Not found in plant cells. Influence cell shape, move chromosomes during reproductive division, and are the internal structure for flagellum.

— *Secretory Vesicles*: Membrane enclosed sack created at the golgi complex. These structures contain cell secretions, like hormones and neurotransmitters. The secretory vesicles are then transported to the cell surface where they are release to the environment outside the cell.

— *Vacuole*: Voids within the cytoplasm. Quite large in plant cells. Used to store water and waste products.

— *Cytoplasm*: Semifluid mixture that occupies most of the cell's interior. Contains sugars, amino acids, and proteins. Also, contains a protein fiber network. Medium in which organelles and other internal structures exist in. Fiber network support the shape of the cell and anchor organelles to fixed positions.

— *Mitochondria*: Elongated structures about 1 to 3 micrometers long. Resemble aerobic bacteria. Structure which converts sugar into energy through oxidation.

— *Chloroplasts*: Elongated structures with vesicles containing chlorophyll. Site of photosynthesis.

Cells can also be classified according to how they obtain their energy. Some cells have the ability to use light or chemical energy found in the outside environment to manufacture their own sugars, fats, and proteins. We call these types of cells autotrophs. All 400,000 species of plants and a few species of bacteria use sunlight and the process of photosynthesis to obtain their energy. Some bacteria breakdown molecules found in the environment to release chemical energy to sustain their life. Organisms can also obtain their energy by consuming other organisms. These organisms are called heterotrophs. Heterotrophs include most types of bacteria and all of the animal and fungi species.

Some organisms consist of just one cell (bacteria, algae, and protozoa). However, most organisms are multi-cellular. Within multi-cellular organisms, groups of cells can become specialized to carry out a certain function. We call these functional groups of cells a tissue. An organ is a structure composed of several different types of tissues. Organs also have a specific structure and a particular function.

Energy Measurement

Now we able to measure and quantify the energy, using a standard set of units. Worldwide, two systems of units of measurement are commoly used today: the Metric System (Systeme International) and the British System. The units of energy described in these systems are derived from a technical definition of energy used by physicists. This definition suggests that energy can be represented by the following simple equation:

$$\text{Work} = \text{Force} \times \text{Distance}$$

Physicists view, energy as the ability to do work. However, they define work as a force applied to some form of matter (object) multiplied by the distance that this object travels. Physicists commonly describe force with a unit of measurement known as a newton (after Sir Isaac Newton). A newton is equal to the force needed to accelerate (move) a mass weighting one kilogram one meter in one second in a vacuum with no friction. The work or energy required to move an object with the force of one newton over a distance of one meter is called a joule.

So far we have learned that energy can take on many forms. One important form of energy, relative to life on

Earth, is kinetic energy. Simply defined, kinetic energy is the energy of motion. The amount of kinetic energy that a body possesses is dependent on the speed of its motion and its mass. At the atomic scale, the kinetic energy of atoms and molecules is sometimes referred to as heat energy. Kinetic energy is also related to the concept of temperature.

Temperature is defined as the measure of the average speed of atoms and molecules. The higher the temperature, the faster these particles of matter move. At a temperature of - 273.15° Celsius (absolute zero) all atomic motion stops. Heat is often defined as energy in the process of being transferred from one object to another because of the temperature difference between them. In the atmosphere, heat is commonly transferred by conduction, convection, advection, and radiation. Some other important definitions related to energy, temperature, and heat are:

— *Heat Capacity*: of a substance is the ratio of the amount of heat energy absorbed by that substance to its corresponding temperature rise.

— *Specific Heat*: is the heat capacity of a unit mass of a substance or heat needed to raise the temperature of 1 gram (g) of a substance 1 degree Celsius.

— *Sensible Heat*: is heat that can be measured by a thermometer, and thus sensed by humans. Several different scales of measurement exist for measuring sensible heat. The most common are: Celsius scale, Fahrenheit scale, and the Kelvin scale.

— *Latent Heat*: is the energy required to change a substance to a higher state of matter. This same energy is released from the substance when the change of state is reversed. The diagram below describes the various exchanges of heat involved with 1 gram of water.

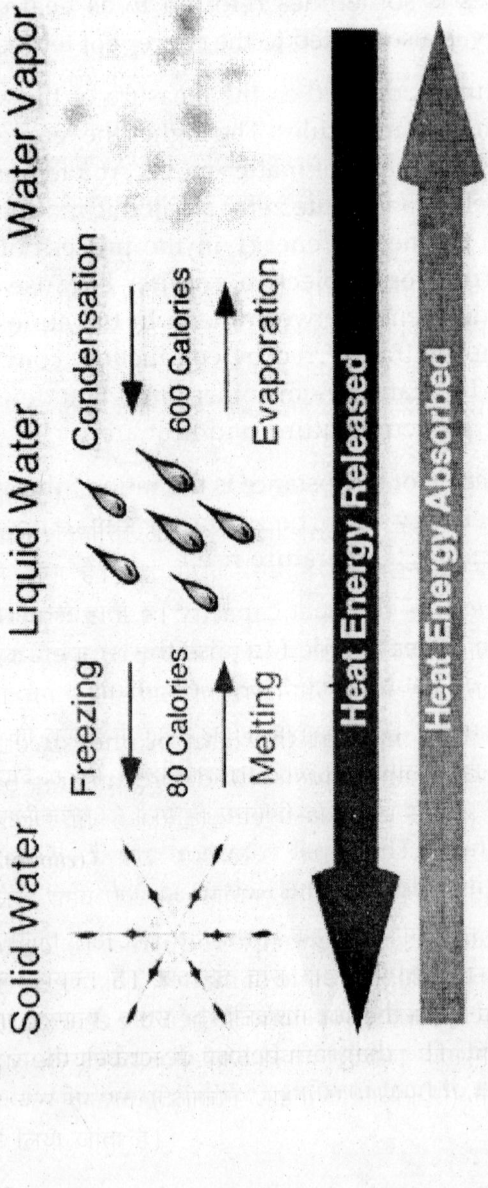

Figure 4: Latent heat exchanges of energy involved with the phase changes of water.

The highest values of flux or flow occur near the subtropical oceans where high temperatures and a plentiful supply of water encourage the evaporation of water. Negative values of latent heat flux indicate a net release of latent energy back into the environment because of the condensation or freezing of water. Values of latent heat flux are generally low over landmasses because of a limited supply of water at the ground surface. The capture and use of energy in living systems is dominated by two processes: photosynthesis and respiration. Through these two processes living organisms are able to capture and use all of the energy they require for their activities.

Photosynthesis

Plants can capture the electromagnetic energy from the sun by a chemical process called photosynthesis. This chemical reaction can be described by the following simple equation:

$$6CO_2 + 6H_2O + \text{light energy} >>> C_6H_{12}O_6 + 6O_2$$

The product of photosynthesis is the carbohydrate glucose and oxygen which is released into the atmosphere. All of the sugar glucose is produced in the specialized photosynthetic cells of plants and some other organisms. Glucose is produced by chemically combining carbon dioxide and water with sunlight. This chemical reaction is catalyzed by chlorophyll acting in concert with other pigment, lipid, sugars, protein, and nucleic acid molecules.

Sugars created in photosynthesis can be later converted by the plant to starch for storage, or it can be combined with other sugar molecules to form specialized carbohydrates such as cellulose, or it can be combined with other nutrients such as nitrogen, phosphorus, and sulfur, to build complex molecules such as proteins and nucleic acids. Because all the energy fixed by a plant is converted into sugar, it is theoretically possible to determine a plant's energy uptake by measuring the amount of sugar produced. This quantity is called gross primary productivity.

Measurements of the buildup of sugar in the plant reflect gross primary productivity less respiration, or net primary productivity. In general, animals cannot produce their own energy via photosynthesis. Instead, they capture their energy by the consumption and assimilation of the biomass of plants or other animals. Thus, animals get the energy they need for maintenance of their bodies tissues, growth, and reproduction indirectly from photosynthetic organisms.

Respiration

The oxidation of sugar by organisms is called respiration. This process occurs in both plants and animals. In most organisms, respiration releases the energy required for all metabolic processes. This chemical reaction can be described by the following simple equation:

$$C_6H_{12}O_6 + 6O_2 >>> 6CO_2 + 6H_2O + \text{released energy}$$

One of the products of respiration is energy, which is released via the chemical decomposition of glucose. Other products of this chemical reaction are carbon dioxide (CO_2) and water (H_2O).

Thermodynamics

The field of thermodynamics studies the behavior of energy flow in natural systems. From this study, a number of physical laws have been established. The laws of thermodynamics describe some of the fundamental truths of thermodynamics observed in our Universe. Understanding these laws is important to students of Physical Geography because many of the processes studied involve the flow of energy.

First Law of Thermodynamics

The first law of thermodynamics is often called the Law of Conservation of Energy. This law suggests that energy can be transferred from one system to another in many forms. However, it can not be created nor destroyed. Thus, the total amount of energy available in the Universe is constant. Einstein's famous equation (written below) describes the relationship between energy and matter:

$$E = MC^2$$

In the equation above, energy (E) is equal to matter (M) times the square of a constant (C). Einstein suggested that energy and matter are interchangeable. His equation also suggests that the quantity of energy and matter in the Universe is fixed.

Second Law of Thermodynamics

Heat can never pass spontaneously from a colder to a hotter body. As a result of this fact, natural processes that involve energy transfer must have one direction, and all natural processes are irreversible. This law also predicts that the entropy of an isolated system always increases with time. Entropy is the measure of the disorder or randomness of energy and matter in a system. Because of the second law of thermodynamics both energy and matter in the Universe are becoming less useful as time goes on. Perfect order in the Universe occurred the instance after the Big Bang when energy and matter and all of the forces of the Universe were unified.

Third Law of Thermodynamics

The third law of thermodynamics states that if all the thermal motion of molecules (kinetic energy) could be removed, a state called absolute zero would occur. Absolute zero results in a temperature of 0 Kelvin or -273.15° Celsius.

Absolute Zero = 0 Kelvin = -273.15° Celsius

The Universe will attain absolute zero when all energy and matter is randomly distributed across space. The current temperature of empty space in the Universe is about 2.7 Kelvins.

Radiation

All objects above the temperature of absolute zero (-273.15° Celsius) radiate energy to their surrounding environment. This energy, or radiation, is emitted as electromagnetic waves that travel at the speed of light. Many different types of radiation have been identified. Each of these types is defined by its wavelength. The wavelength of electromagnetic radiation can vary from being infinitely short to infinitely long.

Visible light is a form of electromagnetic radiation that can be perceived by our eyes. Light has a wavelength of between 0.40 to 0.71 micrometers (μm). Solar radiation spans a spectrum from approximately 0.1 to 4.0 micrometers. The band from 0.1 to 0.4 micrometers is called ultraviolet radiation. About 7 % of the sun's emission is in this wavelength band. About 48 % of the sun's radiation falls in the region between 0.71 to 4.0 micrometers. This band is called the near (0.71 to 1.5 micrometers) and far infrared (1.5 to 4.0 micrometers).

The amount of electromagnetic radiation emitted by a body is directly related to its temperature. If the body is a perfect emitter (black body), the amount of radiation given off is proportional to the 4th power of its temperature as measured in Kelvin units. This natural phenomenon is described by the Stephan-Boltzmann Law. The following simple equation describes this law mathematically:

$$E^* = \sigma T^4$$

where σ (sigma) = $5.67 \times 10^{-8}\ Wm^{-2}\ K^{-4}$

and T is the temperature in Kelvin

According to the Stephan-Boltzmann equation, a small increase in the temperature of a radiating body results in a large amount of additional radiation being emitted. In general, good emitters of radiation are also good absorbers of radiation at specific wavelength bands. This is especially true of gases and is responsible for the Earth's greenhouse effect. Likewise, weak emitters of radiation are also weak absorbers of radiation at specific wavelength bands. This fact is referred to as Kirchhoff's Law. Some objects in nature have almost completely perfect abilities to absorb and emit radiation. We call these objects black bodies. The radiation characteristics of the sun and the Earth are very close to being black bodies.

The wavelength of maximum emission of any body is inversely proportional to its absolute temperature. Thus, the higher the temperature, the shorter the wavelength of maximum emission. This phenomenon is often called Wien's Law. The following equation describes this law:

$$\lambda_{max} = C/T$$

where C is a constant equal to 2897
and T is the temperature in Kelvin

Wien's law suggests that as the temperature of a body increases, the wavelength of maximum emission becomes smaller. According to the above equation the wavelength of maximum emission for the sun (5800 Kelvins) is about 0.5 micrometers, while the wavelength of maximum emission for the Earth (288 Kelvins) is approximately 10.0 micrometers.

The amount of radiation passing through a specific area is inversely proportional to the square of the distance of that area from the energy source. This phenomenon is called the Inverse Square Law. Using this law we can model the effect that distance traveled has on the intensity of emitted radiation from a body like the sun.

Mathematically, the Inverse Square Law is described by the equation:

$$\text{Intensity} = I/d^2$$

where I is the intensity of the radiation at 1d and d is the distance traveled.

Let us try this equation out. For example, what would be the intensity of emitted radiation traveling two units of distance if the intensity at 1d = 90,000 units?

Intensity = $90,000/2^2$

Intensity = 90,000/4

Intensity = 22,500 units

Note that the decrease in intensity with distance is not linear when graphed (Figure 5)

Almost all of the energy that drives the various systems (climate systems, ecosystems, hydrologic systems, etc.) found on the Earth originates from the sun. Solar energy is created at the core of the sun when hydrogen atoms are fused into helium by nuclear fusion. For each second of this nuclear process, 700 million tons of hydrogen are converted into 695 million tons of helium. The remaining 5 million tons are turned into electromagnetic energy that radiates from the sun's surface out into space.

The radiative surface of the sun, or photosphere, has an average temperature of about 5800 Kelvins. Most of the electromagnetic radiation emitted from the sun's surface lies in the visible band centered at 0.5 μm. The total quantity of energy emitted from the sun's surface is approximately 63,000,000 Watts per square meter (Wm^{-2}).

The energy emitted by the sun passes through space until it is intercepted by planets and other celestial objects. The intensity of solar radiation striking these objects is determined by a physical law known as the Inverse Square Law. This law merely states that the intensity of the radiation emitted from the sun varies with the squared distance from the source.

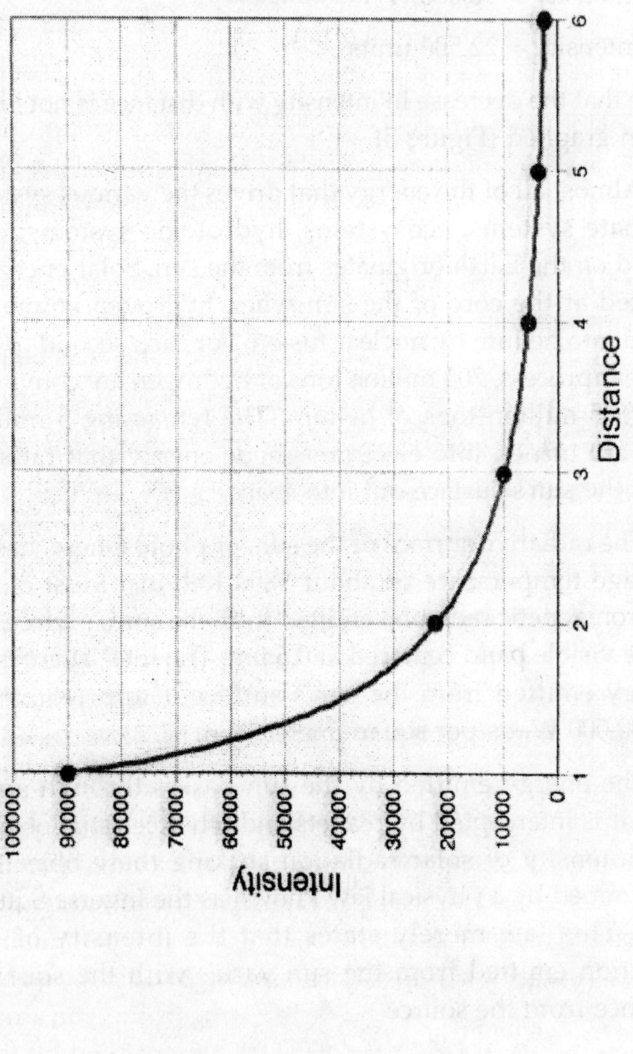

Figure 5: Reduction in intensity of radiation with distance traveled.

As a result of this law, if the intensity of radiation at a given distance is one unit, at twice the distance the intensity will become only one-quarter.

At three times the distance, the intensity will become only one-ninth of its original intensity at a distance of one unit, and so on. Given the amount of energy radiated by the sun and the average Earth-sun distance of 149.5 million kilometers, the amount of radiation intercepted by the outer limits of the atmosphere can be calculated to be around 1370 Wm-2. For general purposes, the energy output of the sun can be considered constant. This of course is not entirely true. Scientists have shown that the output of the sun is temporally variable. Some researchers have also suggested that the increase in the average global temperature over the last century may have been solar in origin. This statement, however, is difficult to prove because accurate data on solar output of radiation only goes back to about 1978.

Earth-Sun Geometry

The term Earth rotation refers to the spinning of the Earth on its axis. One rotation takes exactly twenty-four hours and is called a mean solar day. If you could look down at the Earth's North Pole from space you would notice that the direction of rotation is counterclockwise. The opposite is true if you viewed the Earth from the South Pole.

The orbit of the Earth around the sun is called Earth revolution. This celestial motion takes 365 1/4 days to complete one cycle. Further, the Earth's orbit around the sun is not circular, but elliptical (see Figure 6). An elliptical orbit causes the Earth's distance from the sun to vary annually. However, this phenomenon does not cause the seasons! This annual variation in the distance from the sun does influence the amount of solar radiation intercepted by the Earth by approximately 6 %.

On January 3rd, perihelion, the Earth is closest to the sun (147.5 million kilometers). The Earth is farthest from the sun on July 4th, or aphelion. The average distance of the Earth from the sun over a one year period is 150 million kilometers. The Earth's axis is not perpendicular to the plane of the ecliptic, but inclined at a fixed angle of 23.5°. Moreover, the northern end of the Earth's axis always points to the same place in space (North Star).

During the summer solstice, all locations North of the equator have day lengths greater than twelve hours, while all locations South of the equator have day lengths less than twelve hours. On December 21 or 22, the winter solstice, the Earth is positioned so that the South Pole is leaning 23.5° toward the sun. During the winter solstice, all locations North of the equator have day lengths less than twelve hours, while all locations South of the equator have day lengths greater than twelve hours.

During the summer solstice, the area above the Arctic Circle is experiencing 24 hours of daylight because the North Pole is tilted 23.5° toward the sun. The Arctic Circle experiences 24 hours of night when the North Pole is tilted 23.5° away from the sun in the winter solstice. During the two equinoxes, the circle of illumination cuts through the polar axis and all locations on the Earth experience 12 hours of day and night.

On September 22 or 23, the autumnal equinox, neither pole is tilted toward the sun. March 20 or 21 marks the arrival of the spring or vernal equinox when once again the poles are not tilted toward the sun. Day lengths on both of these days, regardless of latitude, are exactly 12 hours.

The annual change in the relative position of the Earth's axis in relationship to the sun causes the height of the sun (solar altitude) to vary in our skies. The total variation in

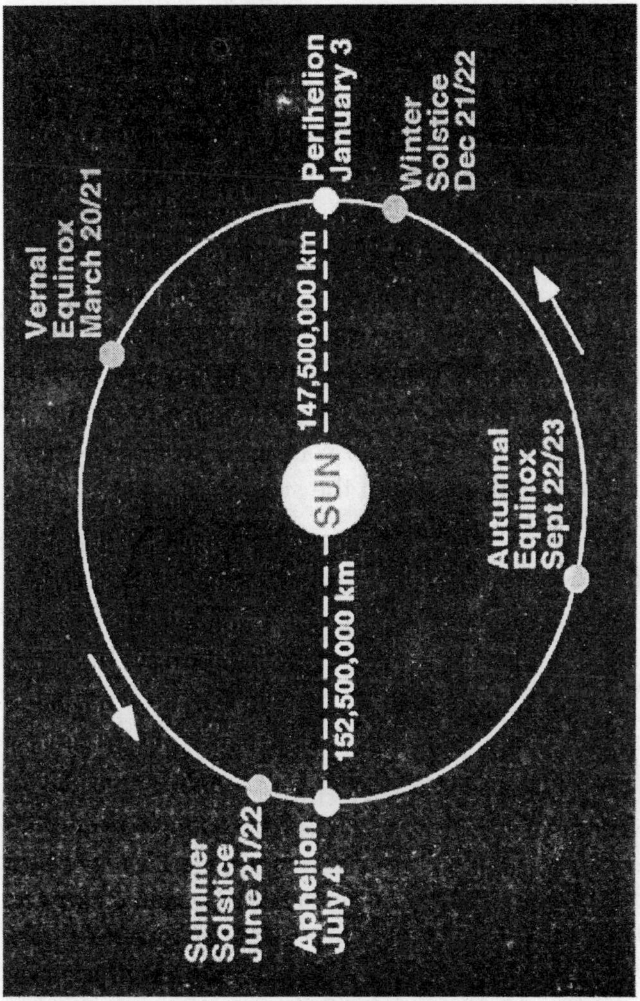

Figure 6: Position of the equinoxes, solstices, aphelion, and perihelion on the Earth's orbit.

maximum solar altitude for any location on the Earth over a one year period is 47° (2 x 23.5 = 47). For example, at 50° North maximum solar altitude varies from 63.5° on the summer solstice to 16.5° on the winter solstice. Maximum solar height at the equator goes from 66.5° above the northern end of the horizon during the summer solstice, to directly overhead on the fall equinox, and then down to 66.5° above the southern end of the horizon during the summer solstice.

The location on the Earth where the sun is directly overhead at solar noon is known as the subsolar point. The subsolar point occurs on the equator during the equinoxes. During the summer solstice, the subsolar point moves to the Tropic of Cancer because at this time the North Pole is tilted 23.5° toward the sun. The subsolar point is located at the Tropic of Capricorn on the winter solstice. On this date, the South Pole is now tilted toward the sun.

The altitude of the sun at solar noon can also be calculated with the following simple equation:

$$\text{Altitude } A = 90 - \text{Latitude } L +/- \text{Declination } D$$

In this equation, L is the latitude of the location in degrees and D is the declination. The equation is simplified to A = 90 - L if sun angle determinations are being made for either equinox date. If the sun angle determination is for a solstice date, declination (D) is added to latitude (L) if the location is experiencing summer (northern latitudes = summer solstice; southern latitudes = winter solstice) and subtracted from latitude (L) if the location is experiencing winter (northern latitudes = winter solstice; southern latitudes = summer solstice). All answers from this equation are given relative to True North for southern latitudes and True South for northern latitudes.

For our purposes only the declinations of the two solstices and two equinoxes are important. These values are: Summer Solstice D=23.5, Winter Solstice D=23.5, Autumnal Equinox D=0, and Vernal Equinox D=0. When using the above equation in tropical latitudes, sun altitude values greater than 90° may occur for some calculations. When this occurs, the noonday sun is actually behind you when looking towards the equator. Under these circumstances, sun altitude should be recalculated as follows:

Altitude A = 90 - (originally calculated Altitude A - 90)

Earth-Sun Relationships

Earth's seasons are controlled by changes in the duration and intensity of solar radiation or insolation. Both of these factors are in turn governed by the annual change in the position of the Earth's axis relative to the sun. Yearly changes in the position of the Earth's axis cause the location of the sun to wander 47° across our skies. Changes in the location of the sun have a direct effect on the intensity of solar radiation.

The intensity of solar radiation is largely a function of the angle of incidence, the angle at which the sun's rays strike the Earth's surface. If the sun is positioned directly overhead or 90° from the horizon, the incoming insolation strikes the surface of the Earth at right angles and is most intense. If the sun is 45° above the horizon, the incoming insolation strikes the Earth's surface at an angle. This causes the rays to be spread out over a larger surface area reducing the intensity of the radiation.

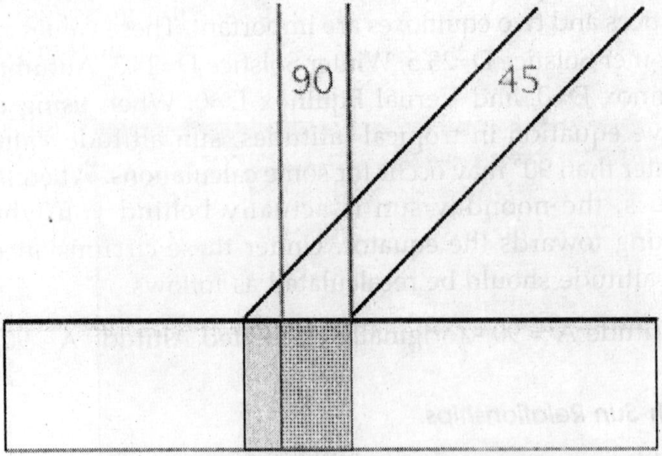

Figure 7: Effect of angle on the area that intercepts an incoming beam of radiation.

Figure 7 models the effect of changing the angle of incidence from 90 to 45°. As illustrated, the lower sun angle (45°) causes the radiation to be received over a much larger surface area. This surface area is approximately 40 % greater than the area covered by an angle of 90°. The lower angle also reduces the intensity of the incoming rays by 30 %.We can also model the effect the angle of incidence has on insolation intensity with the following simple equation:

$$\text{Intensity} = \text{SIN (A)}$$

where, A is the angle of incidence and SIN is the sine function found on most calculators. Using this equation we can determine that an angle of 90° gives us a value of 1.00 or 100 % (1.00 x 100). Let us compare this maximum value with values determined for other angles of incidence. Note the answers are expressed as a percentage of the potential maximum value.

SIN 80 = 0.98 or 98%

SIN 70 = 0.94 or 94%

SIN 60 = 0.87 or 87%

SIN 50 = 0.77 or 77%

SIN 40 = 0.64 or 64%

SIN 30 = 0.50 or 50%

SIN 20 = 0.34 or 34%

SIN 10 = 0.17 or 17%

SIN 0 = 0.00 or 0%

The yearly changes in the position of the Earth's axis relative to the plane of the ecliptic also causes seasonal variations in day length to all locations outside of the equator. Longest days occur during the summer solstice for locations north of the equator and on the winter solstice for locations in the Southern Hemisphere. The equator experiences equal day and night on every day of the year. Day and night is also of equal length for all Earth locations on the autumnal and vernal equinoxes.

Figure 8 below describes the potential insolation available for the equator and several locations in the Northern Hemisphere over a one-year period. The values plotted on this graph take into account the combined effects of angle of incidence and day length duration. Locations at the equator show the least amount of variation in insolation over a one-year period.

These slight changes in insolation result only from the annual changes in the altitude of the sun above the horizon,

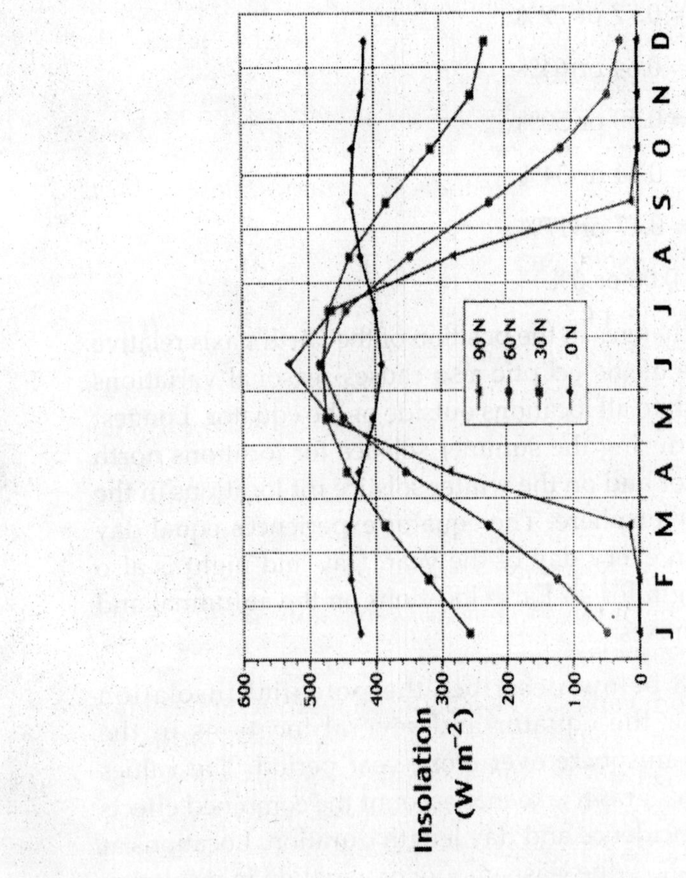

Insolation
(W m-2)
600
500
400
300
200
100
0
J
F
M
A
M
J
J
A
S
O
N
D
Month
90 N
60 N
30 N
0 N

as the duration of daylight at the equator is always 12 hours. The peaks in insolation intensity correspond to the two equinoxes when the sun is directly overhead. The two annual minimums of insolation occur on the solstices when the maximum height of the sun above the horizon reaches an angle of 66.5°.

The most extreme variations in insolation received in the Northern Hemisphere occur at 90° North. During the summer solstice this location receives more potential incoming solar radiation than any other location graphed. At this time the sun never sets. In fact, it remains at an altitude of 23.5° above the horizon for the whole day. From September 22 (autumnal equinox) to March 21, (vernal equinox) no insolation is received at 90° North.

During this period the sun slips below the horizon as the northern axis of the Earth becomes tilted away from the sun. The annual insolation curve for locations at 60° North best approximates the seasonal changes in solar radiation intensity perceived at our latitude. Maximum values of insolation are received at the summer solstice when day length and angle of incidence are at their maximum. During the summer solstice day length is 18 hours and 27 minutes and the angle of the sun reaches a maximum value of 53.5° above the horizon.

Minimum values of insolation are received at the winter solstice when day length and angle of incidence are at their minimum. During the winter solstice day length is only 5 hours and 33 minutes and the angle of the sun reaches a lowest value of 6.5° above the horizon.

4

Atmosphere

Nitrogen and oxygen are the main components of the atmosphere by volume. Together these two gases make up approximately 99 % of the dry atmosphere. Both of these gases have very important associations with life. Nitrogen is removed from the atmosphere and deposited at the Earth's surface mainly by specialized nitrogen fixing bacteria, and by way of lightning through precipitation.

The addition of this nitrogen to the Earth's surface soils and various water bodies supplies much needed nutrition for plant growth. Nitrogen returns to the atmosphere primarily through biomass combustion and denitrification. Oxygen is exchanged between the atmosphere and life through the processes of photosynthesis and respiration. Photosynthesis produces oxygen when carbon dioxide and water are chemically converted into glucose with the help of sunlight.

Respiration is a the opposite process of photosynthesis. In respiration, oxygen is combined with glucose to chemically release energy for metabolism. The products of this reaction are water and carbon dioxide. Water vapor varies in concentration in the atmosphere both spatially and temporally. The highest concentrations of water vapor are found near the equator over the oceans and tropical rain forests. Cold polar areas and subtropical continental deserts are locations where the volume of water vapor can approach zero percent. Water vapor has several very important functional roles on our planet:

— It redistributes heat energy on the Earth through latent heat energy exchange.

— The condensation of water vapor creates precipitaion that falls to the Earth's surface providing needed fresh water for plants and animals.

— It helps warm the Earth's atmosphere through the greenhouse effect.

The fifth most abundant gas in the atmosphere is carbon dioxide. The volume of this gas has increased by over 25 % in the last three hundred years. This increase is primarily due to human induced burning for fossil fuels, deforestation, and other forms of land-use change. Some scientists believe that this increase is causing global warming through an enhancement of the greenhouse effect. Carbon dioxide is also exchanged between the atmosphere and life through the processes of photosynthesis and respiration.

Methane is a very strong greenhouse gas. Since 1750, methane concentrations in the atmosphere have increased by more than 140 %. The primary sources for the additional methane added to the atmosphere are: rice cultivation; domestic grazing animals; termites; landfills; coal mining; and, oil and gas extraction. Anaerobic conditions associated with rice paddy flooding results in the formation of methane gas. However, an accurate estimate of how much methane is being produced from rice paddies has been difficult to ascertain.

More than 60 % of all rice paddies are found in India and China where scientific data concerning emission rates are unavailable. Nevertheless, scientists believe that the contribution of rice paddies is large because this form of crop production has more than doubled since 1950. Grazing animals release methane to the environment as a result of herbaceous digestion. Some researchers believe the addition of methane from this source has more than quadrupled over the last century.

Termites also release methane through similar processes. Land-use change in the tropics, due to deforestation, ranching, and farming, may be causing termite numbers to expand. If this assumption is correct, the contribution from these insects may be important. Methane is also released from landfills, coal mines, and gas and oil drilling. Landfills produce methane as organic wastes decompose over time. Coal, oil, and natural gas deposits release methane to the atmosphere when these deposits are excavated or drilled. The average concentration of the greenhouse gas nitrous oxide is now increasing at a rate of 0.2 to 0.3 % per year.

Its part in the enhancement of the greenhouse effect is minor relative to the other greenhouse gases already mentioned. However, it does have an important role in the artificial fertilization of ecosystems. In extreme cases, this fertilization can lead to the death of forests, eutrophication of aquatic habitats, and species exclusion. Sources for the increase of nitrous oxide in the atmosphere include: land-use conversion; fossil fuel combustion; biomass burning; and soil fertilization.

Most of the nitrous oxide added to the atmosphere each year comes from deforestation and the conversion of forest, savanna and grassland ecosystems into agricultural fields and rangeland. Both of these processes reduce the amount of nitrogen stored in living vegetation and soil through the decomposition of organic matter. Nitrous oxide is also released into the atmosphere when fossil fuels and biomass are burned. However, the combined contribution to the increase of this gas in the atmosphere is thought to be minor.

The use of nitrate and ammonium fertilizers to enhance plant growth is another source of nitrous oxide. How much is released from this process has been difficult to quantify. Estimates suggest that the contribution from this source represents from 50 % to 0.2 % of nitrous oxide added to the atmosphere annually. Ozone's role in the enhancement of the greenhouse effect has been difficult to determine. Accurate measurements of past long-term (more than 25 years in the past) levels of this gas in the atmosphere are currently unavailable.

Moreover, concentrations of ozone gas are found in two different regions of the Earth's atmosphere. The majority of the ozone (about 97 %) found in the atmosphere is concentrated in the stratosphere at an altitude of 15 to 55 kilometers above the Earth's surface. This stratospheric ozone provides an important service to life on the Earth as it absorbs harmful ultraviolet radiation. In recent years, levels of stratospheric ozone have been decreasing due to the buildup of human created chlorofluorocarbons in the atmosphere.

Since the late 1970s, scientists have noticed the development of severe holes in the ozone layer over Antarctica. Satellite measurements have indicated that the zone from 65° North to 65° South latitude has had a 3 % decrease in stratospheric ozone since 1978. Ozone is also highly concentrated at the Earth's surface in and around cities. Most of this ozone is created as a by product of human created photochemical smog. This buildup of ozone is toxic to organisms living at the Earth's surface.

Layers of Atmosphere

The Earth's atmosphere contains several different layers that can be defined according to air temperature. According to temperature, the atmosphere contains four different layers. The first layer is called the troposphere. The depth of this layer varies from about 8 to 16 kilometers. Greatest depths occur at the tropics where warm temperatures causes vertical expansion of the lower atmosphere. From the tropics to the Earth's polar regions the troposphere becomes gradually thinner.

The depth of this layer at the poles is roughly half as thick when compared to the tropics. Average depth of the troposphere is approximately 11 kilometers. Maximum air temperature also occurs near the Earth's surface in this layer. With increasing height, air temperature drops uniformly with altitude at a rate of approximately 6.5° Celsius per 1000 meters. This phenomenon is commonly called the Environmental Lapse Rate. At an average temperature of -56.5° Celsius, the top of the troposphere is reached. At the upper edge of the troposphere is a narrow transition zone known as the tropopause.

Above the tropopause is the stratosphere. This layer extends from an average altitude of 11 to 50 kilometers above the Earth's surface. This stratosphere contains about 19.9 % of the total mass found in the atmosphere. Very little weather occurs in the stratosphere. Occasionally, the top portions of thunderstorms breach this layer. The lower portion of the stratosphere is also infuenced by the polar jet stream and subtropical jet stream. In the first 9 kilometers of the stratosphere, temperature remains constant with height.

A zone with constant temperature in the atmosphere is called an isothermal layer. From an altitude of 20 to 50 kilometers, temperature increases with an increase in altitude. The higher temperatures found in this region of the stratosphere occurs because of a localized concentration of ozone gas molecules. These molecules absorb ultraviolet sunlight creating heat energy that warms the stratosphere. Ozone is primarily found in the atmosphere at varying concentrations between the altitudes of 10 to 50 kilometers. This layer of ozone is also called the ozone layer.

The ozone layer is important to organisms at the Earth's surface as it protects them from the harmful effects of the sun's ultraviolet radiation. Without the ozone layer life could not exist on the Earth's surface. Separating the mesosphere from the stratosphere is transition zone called the stratopause. In the mesosphere, the atmosphere reaches its coldest temperatures (about -90° Celsius) at a height of approximately 80 kilometers. At the top of the mesosphere is another transition zone known as the mesopause.

The last atmospheric layer has an altitude greater than 80 kilometers and is called the thermosphere. Temperatures in this layer can be as high as 1200° C. These high temperatures are generated from the absorption of intense solar radiation by oxygen molecules (O_2). While these temperatures seem extreme, the amount of heat energy involved is very small. The amount of heat stored in a substance is controlled in part by its mass. The air in the thermosphere is extremely thin with individual gas molecules being separated from each other by large distances.

Consequently, measuring the temperature of thermosphere with a thermometer is a very difficult process. Thermometers measure the temperature of bodies via the movement of heat energy. Normally, this process takes a few minutes for the conductive transfer of kinetic energy from countless molecules in the body of a substance to the expanding liquid inside the thermometer. In the thermosphere, our thermometer would lose more heat energy from radiative emission then what it would gain from making occasional contact with extremely hot gas molecules.

The atmosphere is composed of a mixture of many different gases. This mixture behaves in many ways as if it were a single gas. As a result of this phenomenon, the following generalizations describe important relationships between temperature, pressure, density and volume, that relate to the Earth's atmosphere.

1. When temperature is held constant, the density of a gas is proportional to pressure, and volume is inversely proportional to pressure. Accordingly, an increase in pressure will cause an increase in density of the gas and a decrease in its volume.
2. If volume is kept constant, the pressure of a unit mass of gas is proportional to temperature. If temperature increase so will pressure, assuming no change in the volume of the gas.
3. Holding pressure constant, causes the temperature of a gas to be proportional to volume, and inversely proportional to density. Thus, increasing temperature of a unit mass of gas causes its volume to expand and its density to decrease as long as there is no change in pressure.

These relationships can also be described mathematically by the Ideal Gas Law. Two equations that are commonly used to describe this law are:

Pressure x Volume = Constant x Temperature

and

Pressure = Density × Constant × Temperature

Air is a tangible material substance and as a result has mass. Any object with mass is influenced by the universal force known as gravity. Newton's Law of Universal Gravitation states: any two objects separated in space are attracted to each other by a force proportional to the product of their masses and inversely proportional to the square of the distance between them. On the Earth, gravity can also be expressed as a force of acceleration of about 9.8 meters per second per second.

As a result of this force, the speed of any object falling towards the surface of the Earth accelerates (1st second - 9.8 meters per second, 2nd second - 19.6 meters per second, 3rd second - 29.4 meters per second, and so on.) until terminal velocity is attained. Gravity shapes and influences all atmospheric processes. It causes the density and pressure of air to decrease exponentially as one moves away from the surface of the Earth.

Atmospheric Pressure

Any instrument that measures air pressure is called a barometer. The first measurement of atmospheric pressure began with a simple experiment performed by Evangelista Torricelli in 1643. In his experiment, Torricelli immersed a tube, sealed at one end, into a container of mercury. Atmospheric pressure then forced the mercury up into the tube to a level that was considerably higher than the mercury in the container. Torricelli determined from this experiment that the pressure of the atmosphere is approximately 30 inches or 76 centimeters (one centimeter of mercury is equal to 13.3 millibars). He also noticed that height of the mercury varied with changes in outside weather conditions.

The most common type barometer used in homes is the aneroid barometer. Inside this instrument is a small, flexible metal capsule called an aneroid cell. In the construction of the device, a vacuum is created inside the capsule so that small changes in outside air pressure cause the capsule to expand or contract. The size of the aneroid cell is then calibrated and any change in its volume is transmitted by springs and levers to an indicating arm that points to the corresponding atmospheric pressure.

For climatological and meteorological purposes, standard sea-level pressure is said to be 76.0 cm or 29.92 inches or 1013.2 millibars. Scientists often use the kilopascal (kPa) as their preferred unit for measuring pressure. 1 kilopascal is equal to 10 millibars. Another unit of force sometimes used by scientists to measure atmospheric pressure is the newton. One millibar equals 100 newtons per square meter (N/m2).

Atmospheric Pressure at the Earth's Surface

During the winter months (December to February), areas of high pressure develop over central Asia (Siberian High), off the coast California (Hawaiian High), central North America (Canadian High), over Spain and northwest Africa extending into the subtropical North Atlantic (Azores High), and over the oceans in the Southern Hemisphere at the subtropics. Areas of low pressure occur just south of the Aleutian Islands (Aleutian Low), at the southern tip of Greenland (Iceland Low), and latitudes 50 to 80° South.

During the summer months (June to August), a number of dominant winter pressure systems disappear. Gone are the Siberian High over central Asia and the dominant low pressure systems near the Aleutian Islands and at the southern tip of Greenland. The Hawaiian and Azores High intensify and expand northward into their relative ocean basins. High pressure systems over the subtropical oceans in Southern Hemisphere also intensity and expand to the north. New areas of dominant high pressure develop over Australia and Antarctica (South Polar High).

Regions of low pressure form over central Asia and southwest Asia (Asiatic Low). These pressure systems are responsible for the summer monsoon rains of Asia.

The Ozone Layer

The ozone layer is a region of concentration of the ozone (O_3) molecule in the Earth's atmosphere. The layer sits at an altitude of about 10 to 50 kilometers, with a maximum concentration in the stratosphere at an altitude of approximately 25 kilometers. In recent years, scientists have measured a seasonal thinning of the ozone layer primarily at the South Pole. This phenomenon is being called the ozone hole.

The ozone layer naturally shields Earth's life from the harmful effects of the sun's ultraviolet (UV) radiation. A severe decrease in the concentration of ozone in the ozone layer could lead to the following harmful effects:

— An increase in the incidence of skin cancer (ultraviolet radiation can destroy acids in DNA).

— A large increase in cataracts and sun burning.

— Suppression of immune systems in organisms.

— Adverse impact on crops and animals.

— Reduction in the growth of phytoplankton found in the Earth's oceans.

— Cooling of the Earth's stratosphere and possibly some surface climatic effect.

— Ozone is created naturally in the stratosphere by the combining of atomic oxygen (O) with molecular oxygen (O_2). This process is activated by sunlight. Ozone is destroyed naturally by the absorption of ultraviolet radiation,

$$O_3 + UV >>> O_2 + O,$$

and by the collision of ozone with other atmospheric atoms and molecules.

$$O_3 + O >>> 2O_2$$

$$O_3 + O_3 >>> 3O_2$$

Since the late 1970s, scientists have discovered that stratospheric ozone amounts over Antarctica in springtime (September - November) have decreased by as much as 60 %. Satellite measurements have indicated a 3 % decrease in ozone between 65° North - 65° South since 1978. A reduction of about 3 % per year has been measured for Antarctica where most of the ozone loss is occurring globally.

During the late 1990s, large losses of ozone were recorded above Antarctica year after year in the months of September and August. In some years, spring levels of stratospheric ozone were more than 50 % lower than the levels recorded months prior to the seasonal development of the hole. It appears that human activities are altering the amount of stratospheric O_3. The main agent responsible for this destruction was human-made chlorofluorocarbons or CFCs.

First produced by General Motors Corporation in 1928, CFCs were created as a replacement to the toxic refrigerant ammonia. CFCs have also been used as a propellant in spray cans, cleaner for electronics, sterilant for hospital equipment, and to produce the bubbles in styrofoam. CFCs are cheap to produce and are very stable compounds, lasting up to 200 years in the atmosphere. By 1988, some 320,000 metric tons of CFCs were used worldwide.

In 1987, a number of nations around the world met to begin formulating a global plan, known as the Montreal Protocol, to reduce and eliminate the use of CFCs. Since 1987, the plan has been amended in 1990 and 1992 to quicken the schedule of production and consumption reductions. By 1996, 161 countries were participating in the Protocol. The Montreal Protocol called for a 100 % reduction in the creation and use of CFCs by January 1, 1996 in the world's more developed countries. Less developed countries have until January 1, 2010 to stop their production and consumption of these dangerous chemicals.

CFCs created at the Earth's surface drift slowly upward to the stratosphere where ultraviolet radiation from the sun causes their decomposition and the release of chlorine (Cl). Chlorine in turn attacks the molecules of ozone chemically converting them into oxygen molecules.

$$Cl + O_3 >>> ClO + O_2$$

$$ClO + O >>> Cl + O_2$$

A single chlorine atom removes about 100,000 ozone molecules before it is taken out of operation by other substances. Chlorine is removed from the stratosphere by two chemical reactions:

$$ClO + NO_2 >>> ClONO_2$$

$$CH_4 + Cl >>> HCl + CH_3$$

Normally, these two reactions would quickly neutralize the chlorine released into the stratosphere. However, the presence of polar stratospheric clouds, rich in nitrogen, and sunlight facilitates a series of reactions which prolongs the reactive life of chlorine in the atmosphere. Interestingly, these polar stratospheric clouds require very cold air (approximately -85° Celsius) for their formation. Stratospheric air of this temperature occurs normally every year above Antarctica in the winter and early spring months.

Destruction of the ozone begins in Antarctica in the spring as this region moves from 24 hours of night to 24 hours of day. These clouds are less frequent in the Arctic stratosphere because winter cooling of the air in the stratosphere is less severe. NASA's Earth Probe -Total Ozone Mapping Spectrometer home page has the latest images describing the current status of global stratosphere ozone levels in the atmosphere

Atmospheric Effects on Incoming Solar Radiation

Three atmospheric processes modify the solar radiation passing through our atmosphere destined to the Earth's surface. These processes act on the radiation when it interacts with gases and suspended particles found in the atmosphere. The process of scattering occurs when small particles and gas molecules diffuse part of the incoming solar radiation in random directions without any alteration to the wavelength of the electromagnetic energy (Figure 1). Scattering does, however, reduce the amount of incoming radiation reaching the Earth's surface. A significant proportion of scattered shortwave solar radiation is redirected back to space.

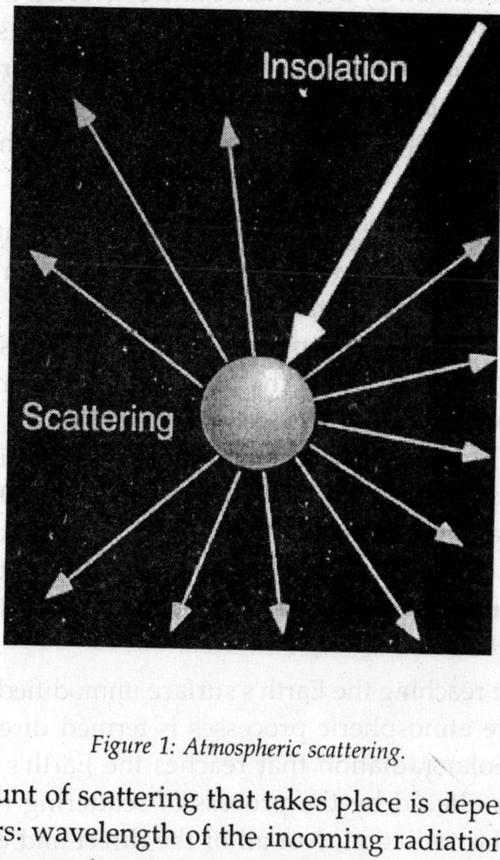

Figure 1: Atmospheric scattering.

The amount of scattering that takes place is dependent on two factors: wavelength of the incoming radiation and the size of the scattering particle or gas molecule. In the Earth's atmosphere, the presence of a large number of particles with a size of about 0.5 microns results in shorter wavelengths being preferentially scattered. This factor also causes our sky to look blue because this color corresponds to those wavelengths that are best diffused. If scattering did not occur in our atmosphere the daylight sky would be black.If intercepted, some gases and particles in the atmosphere have the ability to absorb incoming insolation.

Absorption is defined as a process in which solar radiation is retained by a substance and converted into heat energy. The creation of heat energy also causes the substance to emit its own radiation. In general, the absorption of solar radiation by substances in the Earth's atmosphere results in temperatures that get no higher than 1800° Celsius.

According to Wien's Law, bodies with temperatures at this level or lower would emit their radiation in the longwave band. Further, this emission of radiation is in all directions so a sizable proportion of this energy is lost to space. The final process in the atmosphere that modifies incoming solar radiation is reflection . Reflection is a process where sunlight is redirect by 180° after it strikes an atmospheric particle. This redirection causes a 100 % loss of the insolation. Most of the reflection in our atmosphere occurs in clouds when light is intercepted by particles of liquid and frozen water. The reflectivity of a cloud can range from 40 to 90 %.

Sunlight reaching the Earth's surface unmodified by any of the above atmospheric processes is termed direct solar radiation. Solar radiation that reaches the Earth's surface after it was altered by the process of scattering is called diffused solar radiation. Not all of the direct and diffused radiation available at the Earth's surface is used to do work (photosynthesis, creation of sensible heat, evaporation, etc.). As in the atmosphere, some of the radiation received at the Earth's surface is redirected back to space by reflection. The following image describes the spatial pattern of surface reflectivity as measured for the year 1987.

The reflectivity o. bedo of the Earth's surface varies with the type of material that covers it. For example, fresh snow can reflect up to 95 % of the insolation that reaches it surface. Some other surface type reflectivities are:

— Dry sand 35 to 45 %

— Broadleaf deciduous forest 5 to 10 %

— Needleleaf confierous forest 10 to 20 %

— Grass type vegetation 15 to 25 %

Greenhouse Effect

The greenhouse effect is a naturally occurring process that aids in heating the Earth's surface and atmosphere. It results from the fact that certain atmospheric gases, such as carbon dioxide, water vapor, and methane, are able to change the energy balance of the planet by absorbing longwave radiation emitted from the Earth's surface. Without the greenhouse effect life on this planet would probably not exist as the average temperature of the Earth would be a chilly -18° Celsius, rather than the present 15° Celsius.

As energy from the sun passes through the atmosphere a number of things take place . A portion of the energy (26 % globally) is reflected or scattered back to space by clouds and other atmospheric particles. About 19 % of the energy available is absorbed by clouds, gases (like ozone), and particles in the atmosphere. Of the remaining 55 % of the solar energy passing through the Earth's atmosphere, 4 % is reflected from the surface back to space.

On average, about 51 % of the sun's radiation reaches the surface. This energy is then used in a number of processes, including the heating of the ground surface; the melting of ice and snow and the evaporation of water; and plant photosynthesis. The heating of the ground by sunlight causes the Earth's surface to become a radiator of energy in the longwave band (sometimes called infrared radiation). This emission of energy is generally directed to space. However, only a small portion of this energy actually makes it back to space. The majority of the outgoing infrared radiation is absorbed by the greenhouse gases.

Absorption of longwave radiation by the atmosphere causes additional heat energy to be added to the Earth's atmospheric system. The now warmer atmospheric greenhouse gas molecules begin radiating longwave energy in all directions. Over 90 % of this emission of longwave energy is directed back to the Earth's surface where it once again is absorbed by the surface. The heating of the ground by the longwave radiation causes the ground surface to once again radiate, repeating the cycle described above, again and again, until no more longwave is available for absorption. The amount of heat energy added to the atmosphere by the greenhouse effect is controlled by the concentration of greenhouse gases in the Earth's atmosphere.

All of the major greenhouse gases have increased in concentration since the beginning of the Industrial Revolution (about 1700 AD). As a result of these higher concentrations, scientists predict that the greenhouse effect will be enhanced and the Earth's climate will become warmer. Predicting the amount of warming is accomplished by computer modeling. Computer models suggest that a doubling of the concentration of the main greenhouse gas, carbon dioxide, may raise the average global temperature between 1 and 3° Celsius.

However, the numeric equations of computer models do not accurately simulate the effects of a number of possible negative feedbacks. For example, many of the models cannot properly simulate the negative effects that increased cloud cover would have on the radiation balance of a warmer Earth. Increasing the Earth's temperature would cause the oceans to evaporate greater amounts of water, causing the atmosphere to become cloudier. These extra clouds would then reflect a greater proportion of the sun's energy back to space reducing the amount of solar radiation absorbed by the atmosphere and the Earth's surface. With less solar energy being absorbed at the surface, the effects of an enhanced greenhouse effect may be counteracted.

A number of gases are involved in the human caused enhancement of the greenhouse effect. These gases include: carbon dioxide (CO_2); methane (CH_4); nitrous oxide (N_2O); chlorofluorocarbons (CFxClx); and tropospheric ozone (O_3). Of these gases, the single most important gas is carbon dioxide which accounts for about 55 % of the change in the intensity of the Earth's greenhouse effect. The contributions of the other gases are 25 % for chlorofluorocarbons, 15 % for methane, and 5 % for nitrous oxide. Ozone's contribution to the enhancement of greenhouse effect is still yet to be quantified.

Concentrations of carbon dioxide in the atmosphere are now approaching 360 parts per million. Prior to 1700, levels of carbon dioxide were about 280 parts per million. This increase in carbon dioxide in the atmosphere is primarily due to the activities of humans. Beginning in 1700, societal changes brought about by the Industrial Revolution increased the amount of carbon dioxide entering the atmosphere. The major sources of this gas include fossil fuel combustion for industry, transportation, space heating, electricity generation and cooking; and vegetation changes in natural prairie, woodland, and forested ecosystems.

Emissions from fossil fuel combustion account for about 65 % of the extra carbon dioxide now found in our atmosphere. The remaining 35 % is derived from deforestation and the conversion of prairie, woodland, and forested ecosystems primarily into agricultural systems.

Natural ecosystems can hold 20 to 100 times more carbon dioxide per unit area than agricultural systems. Artificially created chlorofluorocarbons are the strongest greenhouse gas per molecule. However, low concentrations in the atmosphere reduce their overall importance in the enhancement of the greenhouse effect.

Current measurements in the atmosphere indicate that the concentration of these chemicals may soon begin declining because of reduced emissions. Reports of the development of ozone holes over the North and South Poles and a general decline in global stratospheric ozone levels over the last two decades has caused many nations to cutback on their production and use of these chemicals. In 1987, the signing of the Montreal Protocol agreement by forty-six nations established an immediate timetable for the global reduction of chlorofluorocarbons production and use. Since 1750, methane concentrations in the atmosphere have increased by more than 140 %.

The primary sources for the additional methane added to the atmosphere (in order of importance) are rice cultivation, domestic grazing animals, termites, landfills, coal mining, and oil and gas extraction. Anaerobic conditions associated with rice paddy flooding results in the formation of methane gas. However, an accurate estimate of how much methane is being produced from rice paddies has been difficult to obtain. More than 60 % of all rice paddies are found in India and China where scientific data concerning emission rates are unavailable.

Nevertheless, scientists believe that the contribution of rice paddies is large because this form of crop production has more than doubled since 1950. Grazing animals release methane to the environment as a result of herbaceous digestion. Some researchers believe the addition of methane from this source has more than quadrupled over the last century. Termites also release methane through similar processes. Land-use change in the tropics, due to deforestation, ranching, and farming, may be causing termite numbers to expand. If this assumption is correct, the contribution from these insects may be important. Methane is also released from landfills, coal mines, and gas and oil drilling. Landfills produce methane as organic wastes decompose over time.

Coal, oil, and natural gas deposits release methane to the atmosphere when these deposits are excavated or drilled. The average concentration of nitrous oxide in the atmosphere is now increasing at a rate of 0.2 to 0.3 % per year. Sources for this increase include land-use conversion; fossil fuel combustion; biomass burning; and soil fertilization. Most of the nitrous oxide added to the atmosphere each year comes from deforestation and the conversion of forest, savanna and grassland ecosystems into agricultural fields and rangeland. Both of these processes reduce the amount of nitrogen stored in living vegetation and soil through the decomposition of organic matter.

Nitrous oxide is also released into the atmosphere when fossil fuels and biomass are burned. However, the combined contribution of these sources to the increase of this gas in the atmosphere is thought to be minor. The use of nitrate and ammonium fertilizers to enhance plant growth is another source of nitrous oxide. Accurate measurements of how much nitrous oxide is being released from fertilization have been difficult to obtain. Estimates suggest that the contribution from this source may represent from 50 % to 0.2 % of nitrous oxide added to the atmosphere annually. Ozone's role in the enhancement of the greenhouse effect has been difficult to determine scientifically.

Accurate measurements of past long-term (more than 25 years in the past) levels of this gas in the atmosphere are currently unavailable. Concentrations of ozone gas are found in two different regions of the Earth's atmosphere. The majority of the ozone (about 97 %) found in the atmosphere is localized in the stratosphere at an altitude of 15 to 55 kilometers above the Earth's surface. In recent years, the concentration of the stratospheric ozone has been decreasing because of the buildup of chlorofluorocarbons in the atmosphere.

Since the late 1970s, scientists have discovered that total column ozone amounts over Antarctica in the springtime have decreased by as much as 70 %. Satellite measurements have indicated that the zone from 65° North to 65° South latitude has had a 3 % decrease in stratospheric ozone since 1978. Ozone is also highly concentrated at the Earth's surface. Most of this ozone is created as an artificial by product of photochemical smog.

In summary, the greenhouse effect causes the atmosphere to trap more heat energy at the Earth's surface and within the atmosphere by absorbing and re-emitting longwave energy. Of the longwave energy emitted back to space, 90 % is intercepted and absorbed by greenhouse gases. Without the greenhouse effect the Earth's average global temperature would be -18° Celsius, rather than the present 15° Celsius. In the last few centuries, the activities of humans have directly or indirectly caused the concentration of the major greenhouse gases to increase.

Scientists predict that this increase may enhance the greenhouse effect making the planet warmer. Some experts estimate that the Earth's average global temperature has already increased by 0.3 to 0.6° Celsius, since the beginning of this century, because of this enhancement. Predictions of future climates indicate that by the middle of the next century the Earth's global temperature may be 1 to 3° Celsius higher than today.

Ozone Unknown Varies with latitude and altitude in the atmosphere Global levels have generally decreased in the stratosphere and increased near the Earth's surface Created naturally by the action of sunlight on molecular oxygen and artificially through photochemical smog production

Net Radiation and the Planetary Energy Balance

Shortwave radiation from the sun enters the surface-atmosphere system of the Earth and is ultimately returned to space as longwave radiation (because the Earth is cooler than the sun). A basic necessity of this energy interchange is that incoming solar insolation and outgoing radiation be equal in quantity.

The Global Shortwave Radiation Cascade describes the relative amounts (based on 100 units available at the top of the atmosphere) of shortwave radiation partitioned to various atmospheric processes as it passes through the atmosphere.19 units of insolation are absorbed (and therefore transferred into heat energy and longwave radiation) in the atmosphere by the following two processes:

Stratospheric Absorption of the Ultraviolet Radiation by Ozone 2 units; and Tropospheric Absorption of Insolation by Clouds and Aerosols 17 units. 23 units of solar radiation are scattered in the atmosphere subsequently absorbed at the surface as diffused insolation. 28 units of the incoming solar radiation are absorbed at the surface as direct insolation. Total amount of solar insolation absorbed at the surface equals 51 units. The total amount of shortwave radiation absorbed at the surface and in the atmosphere is

Three main losses of solar radiation back to space occur in the Earth's shortwave radiation cascade. 4 units of sunlight are returned to space from surface reflection. Cloud reflection returns another 20 units of solar radiation. Back scattering of sunlight returns 6 units to space. The total loss of shortwave radiation from these processes is 30 units. The term used to describe the combined effect of all of these shortwave losses is Earth albedo.

The Global Longwave Radiation Cascade indicates that energy leaves the Earth's surface through three different processes. 7 units leave the surface as sensible heat. This heat is transferred into the atmosphere by conduction and convection. The melting and evaporation of water at the Earth's surface incorporates 23 units energy into the atmosphere as latent heat. This latent heat is released into the atmosphere when the water condenses or becomes solid. Both of these processes become part of the emission of longwave radiation by the atmosphere and clouds.

The surface of the Earth emits 117 units of longwave radiation. Of this emission only 6 units are directly lost to space. The other 111 units are absorbed by greenhouse gases in the atmosphere and converted into heat energy and then into atmospheric emissions of longwave radiation (the greenhouse effect). The atmosphere emits 160 units of longwave energy. Contributions to this 160 units are from surface emissions of longwave radiation (111 units), latent heat transfer (23 units), sensible heat transfer (7 units), and the absorption of shortwave radiation by atmospheric gases and clouds .

Atmospheric emissions travel in two directions. 64 units of atmospheric emission is lost directly to space. 96 units travel to the Earth's surface where it is absorbed and transferred into heat energy The total amount of energy lost to space in the global longwave radiation cascade is 70 units (surface emission 6 units + atmospheric emission 64 units.) This is the same amount of energy that was added to the Earth's atmosphere and surface by the Global Shortwave Radiation Cascade.

Finally, to balance the surface energy exchanges in this cascade we have to account for 51 units of missing energy [atmosphere and cloud longwave emission (96 units) minus surface longwave emission (117 units) minus latent heat transfer (23 units) minus sensible heat transfer (7 units) = -51 units]. This missing component to the radiation balance is the 51 units of energy absorbed at the Earth's surface as direct and diffused shortwave radiation. The following equations can be used to mathematically model net shortwave radiation balance, net longwave radiation balance, and net radiation balance for the Earth's surface at a single location or for the whole globe for any temporal period:

$$K^* = (K + k)(1 - a)$$

$$L^* = (LD - LU)$$

$$Q^* = (K + k)(1 - a) - LU + LD$$

where, Q^* is surface net radiation (global annual values of $Q^* = 0$, because input equals output, local values can be positive or negative), K^* is surface net shortwave radiation, K is surface direct shortwave radiation, k is diffused shortwave radiation (scattered insolation) at the surface, a is the albedo of surface, L^* is net longwave radiation at the surface, LD is atmospheric counter-radiation (see Greenhouse Effect) directed to the Earth's surface, and LU is longwave radiation lost from the Earth's surface.

Global Heat Balance

The redistribution of energy across the Earth's surface is accomplished primarily through three processes: sensible heat flux, latent heat flux, and surface heat flux into oceans. Sensible heat flux is the process where heat energy is transferred from the Earth's surface to the atmosphere by conduction and convection. This energy is then moved from the tropics to the poles by advection, creating atmospheric circulation. As a result, atmospheric circulation moves warm tropical air to the polar regions and cold air from the poles to the equator.

Latent heat flux moves energy globally when solid and liquid water is converted into vapor. This vapor is often moved by atmospheric circulation vertically and horizontally to cooler locations where it is condensed as rain or is deposited as snow releasing the heat energy stored within it. Finally, large quantities of radiation energy are transferred into the Earth's tropical oceans. The energy enters these water bodies at the surface when absorbed radiation is converted into heat energy. The warmed surface water is then transferred downward into the water column by conduction and convection. Horizontal transfer of this heat energy from the equator to the poles is accomplished by ocean currents. The following equation describes the partitioning of heat energy at the Earth's surface:

$$Q^* = H \text{ (Sensible heat)} + L \text{ (Latent heat)} + S \text{ (Surface heat flux into soil or water)}$$

The actual amount of net radiation being partitioned into each one of these components is a function of the following factors:

— Presence or absence of water in liquid and solid forms at the surface.

— Specific heat of the surface receiving the net radiation.

— Convective and conductive characteristics of the receiving surface.

— Diffusion characteristics of the surface's overlying atmosphere.

— The Concept of Temperature

Temperature and Heat

Temperature and heat are not the same phenomenon. Temperature is a measure of the intensity or degree of hotness in a body. Technically, it is determined by getting the average speed of a body's molecules. Heat is a measure of the quantity of heat energy present in a body. The spatial distribution of temperature in a body determines heat flow. Heat always flows from warmer to colder areas.

The heat held in a object depends not only on its temperature but also its mass. For example, let us compare the heating of two different masses of water. In this example, one mass has a weight of 5 grams, while the other is 25 grams. If the temperature of both masses is raised from 20 to 25° Celsius, the larger mass of water will require five times more heat energy for this increase in temperature. This larger mass would also contain contain 5 times more stored heat energy.

A number of measurement scales have been invented to measure temperature. he most commonly used scale for measuring temperature is the Celsius system. The Celsius scale was developed in 1742 by the Swedish astronomer Anders Celsius. In this system, the melting point of ice was given a value of 0, the boiling point of water is 100, and absolute zero is -273. The Fahrenheit system is a temperature scale that is used exclusively in the United

States. This system was created by German physicist Gabriel Fahrenheit in 1714. In this scale, the melting point of ice has a value of 32, water boils at 212, and absolute zero has a temperature of -460.

The Kelvin scale was proposed by British physicist Lord Kelvin in 1848. This system is often used by scientists because its temperature readings begin at absolute zero and due to the fact that this scale is proportional to the amount of heat energy found in an object. The Kelvin scale assigns a value of 273 for the melting temperature of ice, while the boiling point of water occurs at 373.

A thermometer is a device that is used to measure temperature. Thermometers consist of a sealed hollow glass tube filled with some type of liquid. Thermometers measure temperature by the change in the volume of the liquid as it responds to the addition or loss of heat energy from the environment immediately outside its surface. When heat is added, the liquid inside the thermometer expands..Cooling cause the liquid to contract. Meteorological thermometers are often filled with either alcohol or mercury. Alcohol thermometers are favored in very cold environments because of this liquid's low freezing point (-112° Celsius).

By international agreement, the nations of the world have decided to measure temperature in a similar fashion. This standardization is important for the accurate generation of weather maps and forecasts, both of which depend on having data determined in a uniform way. Weather stations worldwide try to determine minimum and maximum temperatures for each and every day. By averaging these two values, daily mean temperatures are also calculated. Many stations also take temperature readings on the hour.

Temperature measurements are determined by thermometers designed and approved by the World

Meteorological Organization. These instruments are housed in specially designed instrument shelters that allow for the standardization of measurements taken anywhere on the Earth.

Daily and Annual Cycles of Temperature

At the Earth's surface quantities of insolation and net radiation undergo daily cycles of change because the planet rotates on its polar axis once every 24 hours. Insolation is usually the main positive component making up net radiation. Variations in net radiation are primarily responsible for the particular patterns of rising and falling air temperature over a 24 hour period. The following three graphs show hypothetical average curves of insolation, net radiation, and air temperature for a typical land based location at 45° of latitude on the equinoxes and solstices.

Insolation

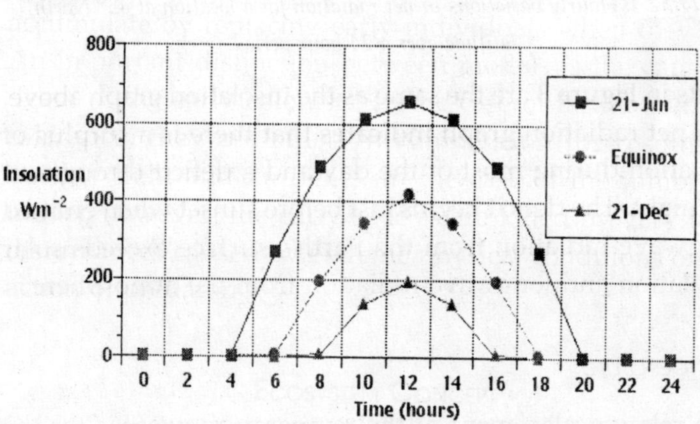

Figure 2: Hourly variations in insolation received for a location at 45° North latitude over a 24 hour period.

In the above graph, shortwave radiation received from the sun is measured in Watts. For all dates, peak reception occurs at solar noon when the sun attains its greatest height above the horizon.

Net Radiation

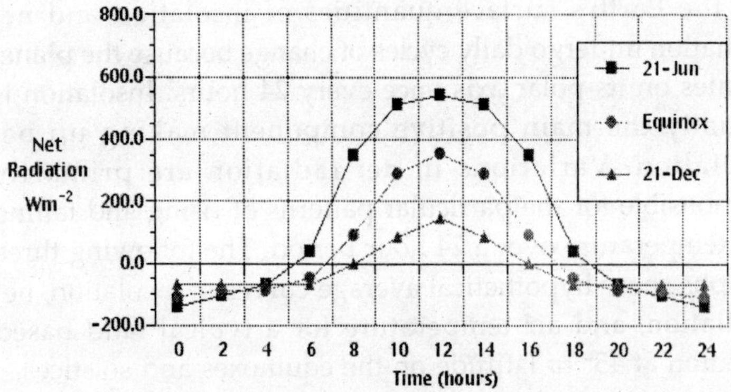

Figure 3: Hourly variations in net radiation for a location at 45° North latitude over a 24 hour period.

Units in Figure 3 are the same as the insolation graph above. The net radiation graph indicates that there is a surplus of radiation during most of the day and a deficit throughout the night. The deficit begins just before sunset when emitted longwave radiation from the Earth's surface exceeds solar insolation and longwave radiation from the atmosphere.

Temperature

The relative placement of the temperature profiles for the various dates correlates to the amount of net radiation available for daily surface absorption and heat generation. The more energy available, the higher up the Y-axis the profile is on the graph. Autumnal equinox (September 21) is warmer than the vernal equinox (March 21) because of the heating that occurred in the previous summer months. For all dates, minimum temperature occurs at sunrise.

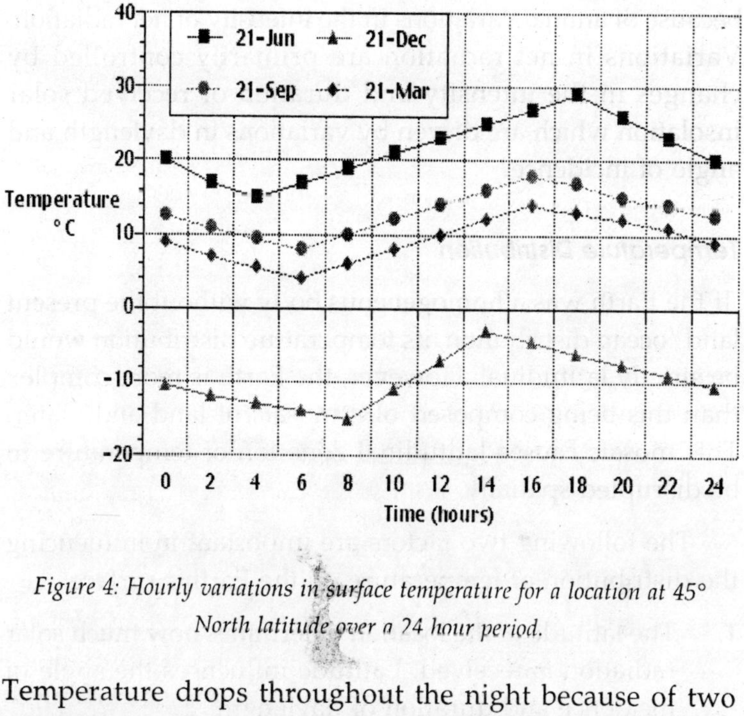

Figure 4: Hourly variations in surface temperature for a location at 45° North latitude over a 24 hour period.

Temperature drops throughout the night because of two processes. First, the Earth's radiation balance at the surface becomes negative after sunset. Thus, the surface of the Earth stops heating up as solar radiation is not being absorbed. Secondly, conduction and convection transport heat energy up into the atmosphere and the warm air that was at the surface is replaced by cooler air from above because of atmospheric mixing. Temperature begins rising as soon as the net radiation budget of the surface becomes positive. Temperature continues to rise from sunrise until sometime after solar noon. After this time, mixing of the Earth's surface by convection causes the surface to cool despite the positive addition of radiation and heat energy.

As the Earth revolves around the sun, locations on the surface may under go seasonal changes in air temperature because of annual variations in the intensity of net radiation. Variations in net radiation are primarily controlled by changes in the intensity and duration of received solar insolation which are driven by variations in daylength and angle of incidence.

Temperature Distribution

If the Earth was a homogeneous body without the present land/ocean distribution, its temperature distribution would be strictly latitudinal. However, the Earth is more complex than this being composed of a mosaic of land and water. This mosaic causes latitudinal zonation of temperature to be disrupted spatially.

The following two factors are important in influencing the distribution of temperature on the Earth's surface:

1. The latitude of the location determines how much solar radiation is received. Latitude influences the angle of incidence and duration of daylength.
2. Surface properties - surfaces with high albedo absorb less incident radiation. In general, land absorbs less insolation that water because of its lighter color. Also, even if two surfaces have the same albedo, a surface's specific heat determines the amount of heat energy required for a specific rise in temperature per unit mass. The specific heat of water is some five times greater than that of rock and the land surface. As a result, water requires the input of large amounts of energy to cause a rise in its temperature.

Mainly because of specific heat, land surfaces behave quite differently from water surfaces. In general, the surface of any extensive deep body of water heats more slowly and cools more slowly than the surface of a large land body. Other factors influencing the way land and water surfaces heat and cool include:

— Solar radiation warms an extensive layer in water, on land just the immediate surface is heated.

— Water is easily mixed by the process of convection.

— Evaporation of water removes energy from water's surface.

During January, much of the terrestrial areas of the Northern Hemisphere are below freezing. Some notable Northern Hemisphere cold-spots include the area around Baffin Island Canada, Greenland, Siberia, and the Plateau of Tibet. Temperatures over oceans tend to be hotter because of the water's ability to hold heat energy.

In the Southern Hemisphere, temperatures over the major landmasses are generally greater than 20° Celsius with localized hot-spots in west-central Australia, the Kalahari Desert in Africa, and the plains of Bolivia, Paraguay, and Argentina. Subtropical oceans are often warmer than landmass areas near the equator. At this latitude, land areas receive less incoming solar radiation because of the daily convective development of cumulus and cumulonimbus clouds.

In the mid-latitudes, oceans are often cooler than landmass areas at similar latitudes. Terrestrial areas are warmer because of the rapid heating of land surfaces under frequently clear skies. Antarctica remains cold and below zero degrees Celsius due to the presence of permanent glacial ice which reflects much of the solar radiation received back to space.

In July, the Northern Hemisphere is experiencing its summer season because the North Pole is now tilted towards the sun. Some conspicuous hot-spots include the south-central United States, Arizona and northwest Mexico, northern Africa, the Middle East, India, Pakistan, and Afghanistan. Temperatures over oceans tend to be relatively cooler because of the land's ability to heat quickly. Two terrestrial areas of cooler temperatures include Greenland and the Plateau of Tibet. In these regions, most of the incoming solar radiation is sent back to space because of the presence of reflective ice and snow.

In the Southern Hemisphere, temperatures over the major landmasses are generally cooler than ocean surfaces at the same latitude. Antarctica is bitterly cold because it is experiencing total darkness. Note that Antarctica is much colder than the Arctic was during its winter season. The Arctic consists mainly of ocean. During the summer, this surface is able to absorb considerable quantities of sunlight which is then converted into heat energy.

The heat stored in the ocean is carried over into the winter season. Antarctica has a surface composed primarily of snow and ice. This surface absorbs only a small amount of the solar radiation during the summer. So it never really heats up. As a result, the amount of heat energy stored into the winter season is minimal.

Wind

Wind can be defined simply as air in motion. This motion can be in any direction, but in most cases the horizontal component of wind flow greatly exceeds the flow that occurs vertically. The speed of wind varies from absolute calm to speeds as high as 380 kilometers per hour. In 1894, strong winds in Nebraska pushed six fully loaded coal cars over 160 kilometers in just over three hours. Over short periods of time surface winds can be quite variable.

Wind develops as a result of spatial differences in atmospheric pressure. Generally, these differences occur because of uneven absorption of solar radiation at the Earth's surface. Wind speed tends to be at its greatest during the daytime when the greatest spatial extremes in atmospheric temperature and pressure exist.

Wind is often described by two characteristics: wind speed and wind direction. Wind speed is the velocity attained by a mass of air traveling horizontally through the atmosphere. Wind speed is often measured with an anemometer in kilometers per hour (kmph), miles per hour (mph), knots, or meters per second (mps). Wind direction is measured as the direction from where a wind comes from. For example, a southerly wind comes from the south and blows to the north. Direction is measured by an instrument called a wind vane. Both of these instruments are positioned in the atmospheric environment at a standard distance of 10 meters above the ground surface.

Wind speed can also be measured without the aid of instruments using the Beaufort wind scale. This descriptive scale was originally developed by Admiral Beaufort of the British Navy in the first decade of the 17th century. The purpose for this system was to allow mariners to determine wind speed from simple observations. The Beaufort system has undergone several modifications to standardize its measurement scale and to allow for its use on land.

Users of this scale look for specific effects of the wind on the environment to determine speed.

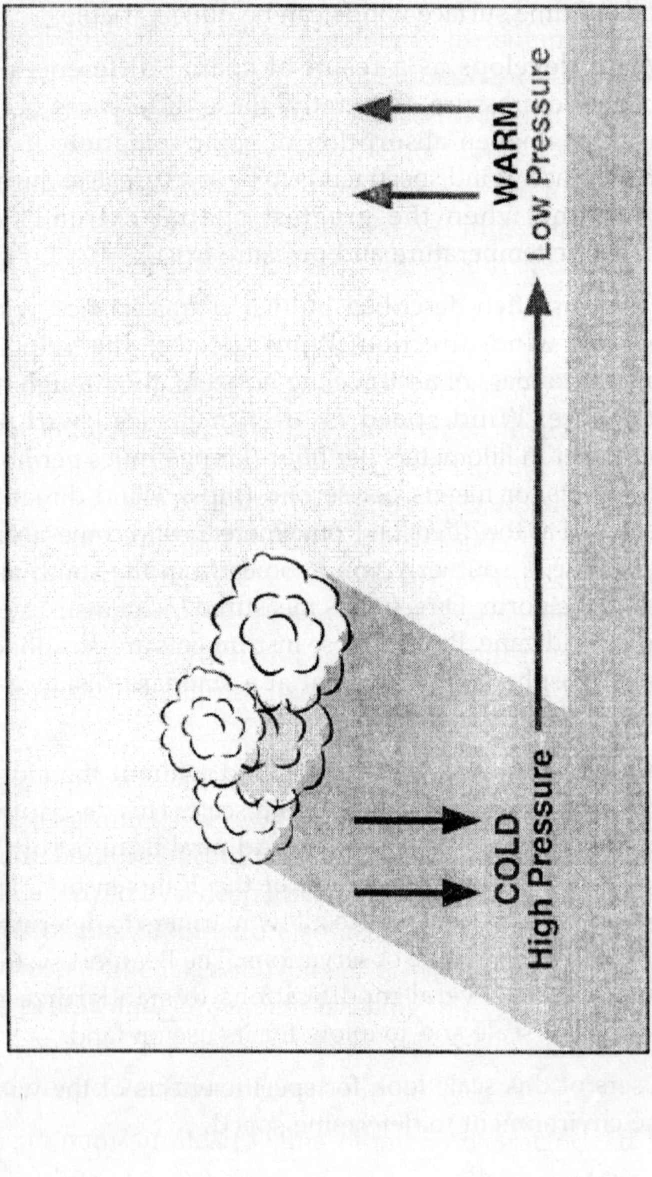

Figure 5: Formation of wind as a result of localized temperature differences.

Winds are named according to the compass direction of their source. Thus, a wind from the north blowing toward the south is called a northerly wind. Figure 6 describes the sixteen principal bearings of wind direction. Most meterological observations report wind direction using one of these sixteen bearings.

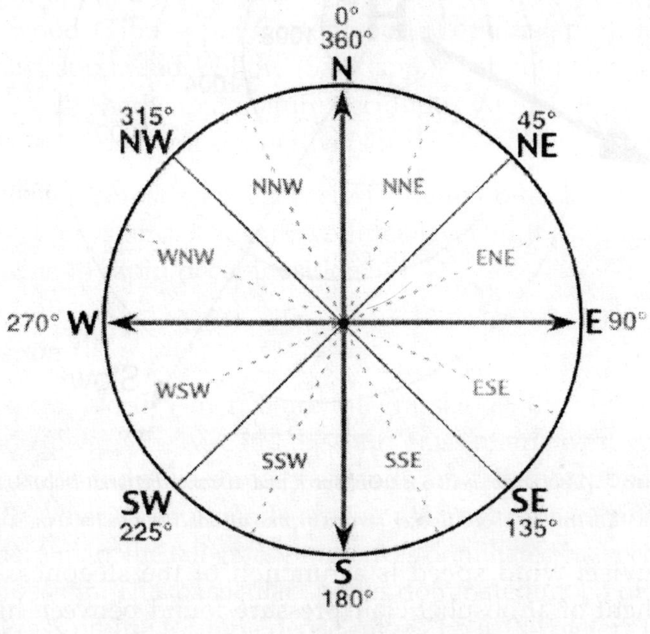

Figure 6: Wind compass describing the sixteen principal bearings used to measure wind direction.

Horizontally, at the Earth's surface wind always blows from areas of high pressure to areas of low pressure (vertically, winds move from areas of low pressure to areas of high pressure), usually at speeds determined by the rate of air pressure change between pressure centers. This situation is comparable to someone skiing down a hill. The skier will of course move from the top of the hill to the bottom of the hill, with the speed of their descent controlled by the gradient or steepness of the slope.

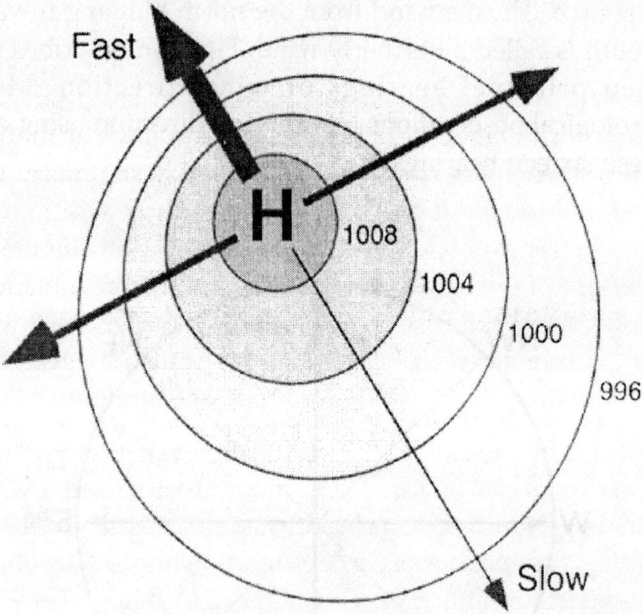

Figure 7: Association between wind speed and distance between isobars. In the illustration above thicker arrows represent relatively faster winds.

Likewise, wind speed is a function of the steepness or gradient of atmospheric air pressure found between high and low pressure systems. When expressed scientifically, pressure change over a unit distance is called pressure gradient force, and the greater this force the faster the winds will blow. On weather maps, pressure is indicated by drawing isolines of pressure, called isobars, at regular 4 millibar intervals (e.g., 996 mb, 1000 mb, 1004 mb, etc.). If the isobars are closely spaced, we can expect the pressure gradient force to be great, and wind speed to be high. In areas where the isobars are spaced widely apart, the pressure gradient is low and light winds normally exist. High speed winds develop in areas where isobars are closer.

To better understand wind we must recognize that it is the result of a limited number of accelerating and decelerating forces, and that the action of these forces is controlled by specific fundamental natural laws. Sir Isaac Newton formulated these laws as several laws of motion. The first law suggests that an object that is stationary will remain stationary, and an object in motion will stay in motion as long as no opposing force is put on the object. As a result of this law, a puck sent in flight from a blade of a hockey stick will remain in motion until friction slows it down or the goalie makes a save. This law also suggests that once in motion an object's path should be straight.

Newton's second law of motion suggests that the force put on an object equals its mass multiplied by the acceleration produced. The term force in this law refers to the total or net effect of all the forces acting on an object. Mathematically, this law is written as:

$$\text{Force} = \text{Mass} \times \text{Acceleration}$$

or

$$\text{Acceleration} = \text{Force}/\text{Mass}$$

From this natural law of motion we can see that the acceleration of an object is directly proportional to the net force pushing or pulling that body and inversely proportional to the mass of the body. Thus, the greater the force created by the movement of a hockey player's stick the faster the puck will travel. This law also suggests that if the player used a larger (more massive) puck more force would have to be applied to it to get it to travel as fast as a puck with less mass.

In the previous lecture we briefly examined one of the forces, pressure gradient force, acting on wind. Let us return to this force and examine it in greater detail and in relation to Newton's laws of motion. We will also examine the effects of three other forces that act on air in motion.

Pressure gradient force is the primary force influencing the formation of wind from local to global scales. This force is determined by the spatial pattern of atmospheric pressure at any given moment in time.

The rotation of the Earth creates another force, termed Coriolis force, which acts upon wind and other objects in motion in very predictable ways. According to Newton's first law of motion, air will remain moving in a straight line unless it is influenced by an unbalancing force. The consequence of Coriolis force opposing pressure gradient acceleration is that the moving air changes direction. Instead of wind blowing directly from high to low pressure, the rotation of the Earth causes wind to be deflected off course.

In the Northern Hemisphere, wind is deflected to the right of its path, while in the Southern Hemisphere it is deflected to the left. The magnitude of the Coriolis force varies with the velocity and the latitude of the object (see Figure 8). Coriolis force is absent at the equator, and its strength increases as one approaches either pole. Furthermore, an increase in wind speed also results in a stronger Coriolis force, and thus in greater deflection of the wind. Coriolis force only acts on air when it has been sent into motion by pressure gradient force. Finally, Coriolis force only influences wind direction and never wind speed.

Centripetal acceleration is the third force that can act on moving air. It acts only on air that is flowing around centers of circulation. Centripetal acceleration is also another force that can influence the direction of wind. Centripetal acceleration creates a force directed at right angles to the flow of the wind and inwards towards the centers of rotation (e.g., low and high pressure centers). This force produces a circular pattern of flow around centers of high and low pressure. Centripetal acceleration is much more important for circulations smaller than the mid-latitude cyclone.

The last force that can influence moving air is frictional deceleration. Friction can exert an influence on wind only after the air is in motion. Frictional drag acts in a direction opposite to the path of motion causing the moving air to decelerate (see Newton's first and second laws of motion). Frictional effects are limited to the lower one kilometer above the Earth's surface.

Air under the influence of both the pressure gradient force and Coriolis force tends to move parallel to isobars in conditions where friction is low (1000 meters above the surface of the Earth) and isobars are straight. Winds of this type are usually called geostrophic winds. Geostrophic winds come about because pressure gradient force and Coriolis force come into balance after the air begins to move.

Buy Ballot's Law states that when you stand with your back to a geostrophic wind in the Northern Hemisphere the center of low pressure will be to your left and the high pressure to your right.

The opposite is true for the Southern Hemisphere. Wind above the Earth's surface does not always travel in straight lines. In many cases winds flow around the curved isobars of a high (anticyclone) or low (cyclone) pressure center.

A wind that blows around curved isobars above the level of friction is called a gradient wind. Gradient winds are slightly more complex than geostrophic winds because they include the action of yet another physical force. This

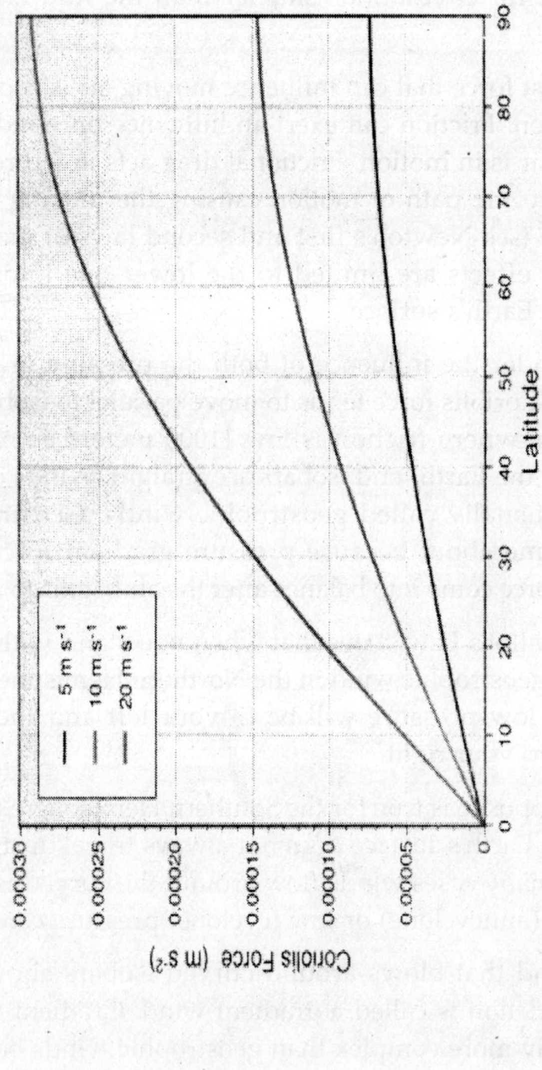

Figure 8: The strength of Coriolis force is influenced by latitude and the speed of the moving object.

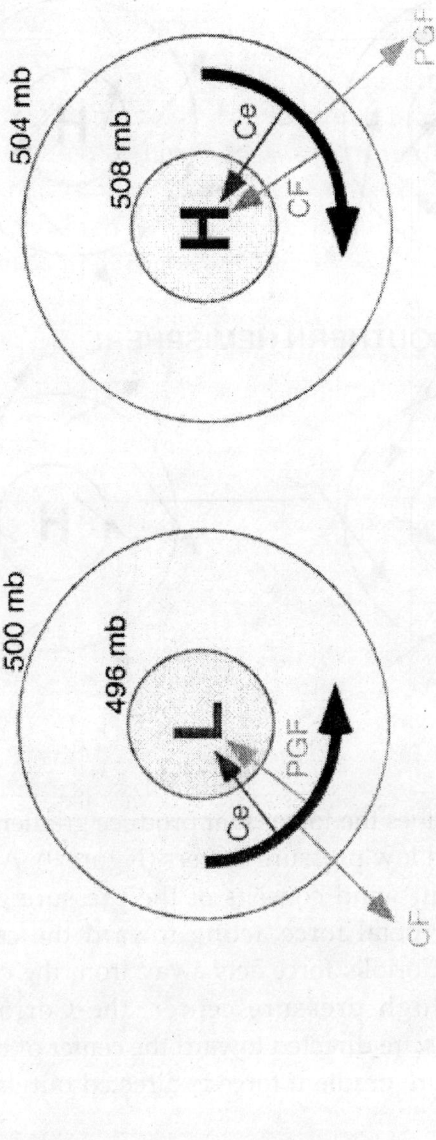

Figure 9: The balance of forces that create a gradient wind in the Northern Hemisphere (PGF = pressure gradient force; CF = Coriolis force; Ce = centripetal force).

force is known as centripetal force and it is always directed toward the center of rotation.

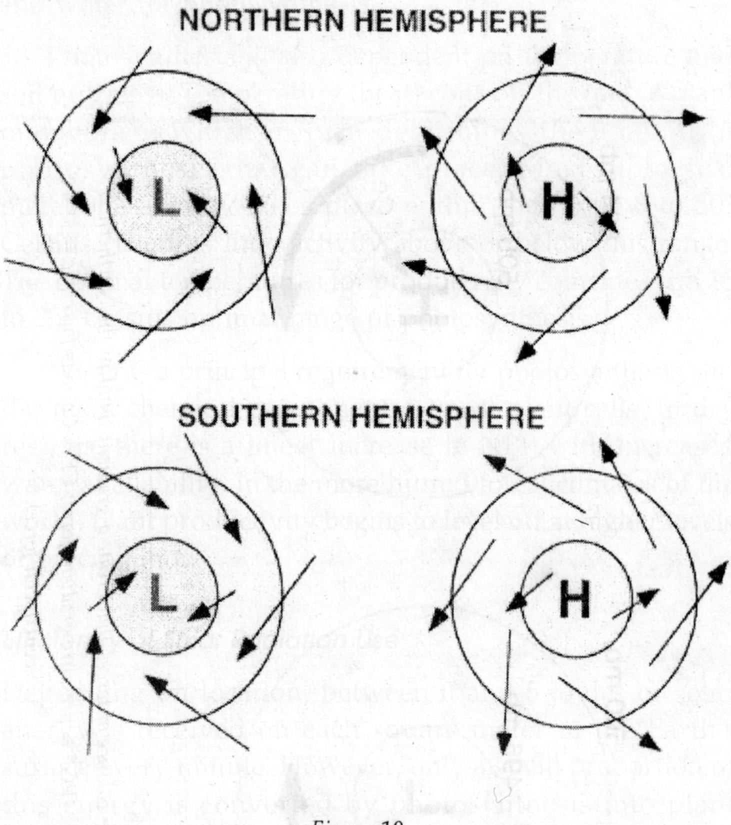

Figure 10.

The figure describes the forces that produce gradient winds around high and low pressure centers (Figure 9). Around a low, the gradient wind consists of the pressure gradient force and centripetal force acting toward the center of rotation, while Coriolis force acts away from the center of the low. In a high pressure center, the Coriolis and centripetal forces are directed toward the center of the high, while the pressure gradient force is directed outward.

Surface winds on a weather map do not blow exactly parallel to the isobars as in geostrophic and gradient winds. Instead, surface winds tend to cross the isobars at an angle varying from 10 to 45°. Close to the Earth's surface, friction reduces the wind speed, which in turn reduces the Coriolis force. As a result, the reduced Coriolis force no longer balances the pressure gradient force, and the wind blows across the isobars toward or away from the pressure center.

The pressure gradient force is now balanced by the sum of the frictional force and the Coriolis force. Thus, we find surface winds blowing counterclockwise and inward into a surface low, and clockwise and out of a surface high in the Northern Hemisphere. In the Southern Hemisphere, the Coriolis force acts to the left rather than the right.

This causes the winds of the Southern Hemisphere to blow clockwise and inward around surface lows, and counterclockwise and outward around surface highs (see Figure 10).

5

Hydrosphere

Physical Properties of Water

We live on a planet that is dominated by water. More than 70 % of the Earth's surface is covered with it. Scientists estimate that the hydrosphere contains about 1.36 billion cubic kilometers of this substance mostly in the form of a liquid that occupies topographic depressions on the Earth. Water is also essential for life. Water is the major constituent of almost all life formost animals and plants contain more than 60 % water by volume. Without water life would probably never have developed on our planet.

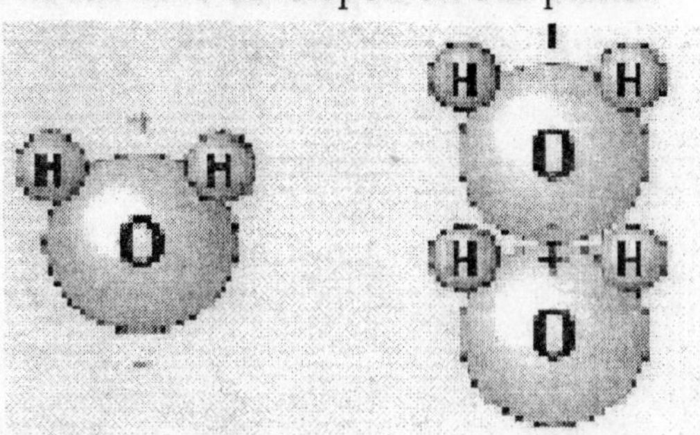

Figure 1: The atomic structure of a water molecule

Water has a very simple atomic structure. This structure consists of two hydrogen atoms bonded to one oxygen atom (Figure 1). The nature of the atomic structure of water causes its molecules to have unique electrochemical properties. The hydrogen side of the water molecule has a slight positive charge (see Figure 1). On the other side of the molecule a negative charge exists. This molecular polarity causes water to be a powerful solvent and is responsible for its strong *ttt*.

In figure 1, the atomic structure of a water molecule consists of two hydrogen (H) atoms joined to one oxygen (O) atom. The unique way in which the hydrogen atoms are attached to the oxygen atom causes one side of the molecule to have a negative charge and the area in the opposite direction to have a positive charge. The resulting polarity of charge causes molecules of water to be attracted to each other forming strong molecular bonds.

Table 1: Density of water at various temperatures

Temperature (degrees Celsius)	Density (grams per cubic centimeter)
0 (solid)	0.9150
0 (liquid)	0.9999
4	1.0000
20	0.9982
40	0.9922
60	0.9832
80	0.9718
100 (gas)	0.0006

When water makes a physical *phase change* its molecules arrange themselves in distinctly different patterns (Figure 2). The pattern taken by water when its is frozen causes its volume to expand and its density to decrease. Expansion of water at freezing allows ice to float on top of liquid water.

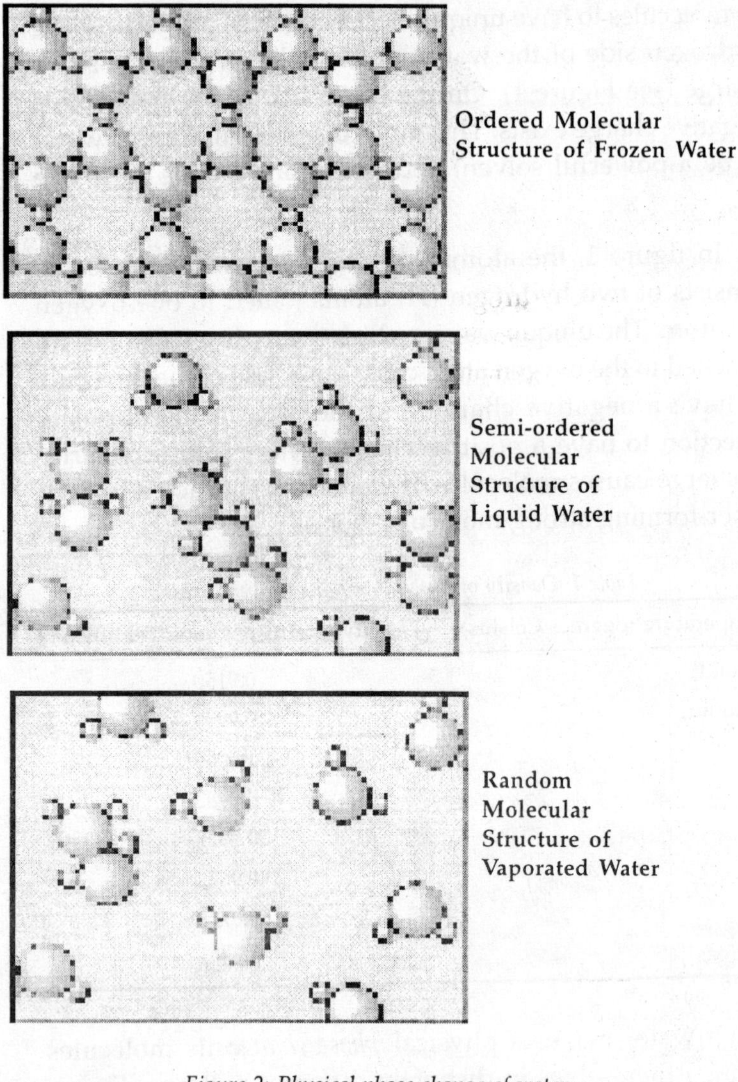

Figure 2: Physical phase change of water

The Hydrologic Cycle

The hydrologic cycle is a conceptual model that describes the storage and movement of water between the biosphere, atmosphere, lithosphere, and the hydrosphere. Water on this planet can be stored in any one of the following reservoirs: atmosphere, oceans, lakes, rivers, soils, glaciers, snowfields, and groundwater.

Water moves from one reservoir to another by way of processes like evaporation, condensation, precipitation, deposition, runoff, infiltration, sublimation, transpiration, melting, and groundwater flow. The oceans supply most of the evaporated water found in the atmosphere. Of this evaporated water, only 91 % of it is returned to the ocean basins by way of precipitation. The remaining 9 % is transported to areas over landmasses where climatological factors induce the formation of precipitation. The resulting imbalance between rates of evaporation and precipitation over land and ocean is corrected by runoff and groundwater flow to the oceans.

The planetary water supply is dominated by the oceans (Table 2). Approximately 97 % of all the water on the Earth is in the oceans. The other 3 % is held as freshwater in glaciers and icecaps, groundwater, lakes, soil, the atmosphere, and within life.

Table 2: Inventory of water at the Earth's surface.

Reservoir	Volume (cubic km × 1,000,000)	Percent of Total
Oceans	1370	97.25
Ice Caps and Glaciers	29	2.05
Groundwater	9.5	0.68
Lakes	0.125	0.01
Soil Moisture	0.065	0.005

Atmosphere	0.013	0.001
Streams and Rivers	0.0017	0.0001
Biosphere	0.0006	0.00004

Water is continually cycled between its various reservoirs. This cycling occurs through the processes of evaporation, condensation, precipitation, deposition, runoff, infiltration, sublimation, transpiration, melting, and groundwater flow. Table 3 describes the approximate residence times of water in the major reservoirs. On average water is renewed in rivers once every 16 days. Water in the atmosphere is completely replaced once every 8 days. Slower rates of replacement occur in large lakes, glaciers, ocean bodies and groundwater. Replacement in these reservoirs can take from hundreds to thousands of years. Some of these resources (especially groundwater) are being used by humans at rates that far exceed their renewal times. This type of resource use is making this type of water effectively nonrenewable.

Table 3: Approximate residence time of water found in various reserviors.

Reservoir	Average Residence Time
Glaciers	40 years
Seasonal Snow Cover	0.4 years
Soil Moisture	0.2 years
Groundwater: Shallow	200 years
Groundwater: Deep	10,000 years
Lakes	100 years
Rivers	0.04 years

Figure 2: The three diagrams illustrate the distinct patterns of molecular arrangement in water when it changes its physical state from ice to water to gas. When water is frozen its molecules arrange themselves in a particular highly

organized rigid geometric pattern that causes the mass of water to expand and to decrease in density. The diagram above shows a slice through a mass of ice that is one molecule wide. In the liquid phase, water molecules arrange themselves into small groups of joined particles. The fact that these arrangements are small allows liquid water to move and flow. Water in the form of a gas is highly charged with energy. This high energy state causes the molecules to be always moving reducing the likelihood of bonds between individual molecules from forming.

Water has several other unique physical properties. These properties are:

— *Water has a high · specific heat.* Specific heat is the amount of energy required to change the temperature of a substance. Because water has a high specific heat, it can absorb large amounts of heat energy before it begins to get hot. It also means that water releases heat energy slowly when situations cause it to cool. Water's high specific heat allows for the moderation of the Earth's climate and helps organisms regulate their body temperature more effectively.

— *Water in a pure state has a neutral · pH.* As a result, pure water is neither · *acidic* nor · *basic.* Water changes its pH when substances are dissolved in it. Rain has a naturally acidic pH of about 5.6 because it contains natural derived carbon dioxide and sulfur dioxide.

— Water· *conducts* heat more easily than any liquid except mercury. This fact causes large bodies of liquid water like lakes and oceans to have essentially a uniform vertical temperature profile.

— Water exists as a liquid over an important range of temperature from 0 - 100° Celsius. This range allows water to remain as a liquid in most places on the Earth.

— Liquid water is a universal solvent. It is able to dissolve a large number of different chemical compounds. This feature also enables water to carry solvent nutrients in *runoff*, *infiltration*, *groundwater flow*, and living organisms.

— *Water has a high surface tension* (Figures 3 and 4). In other words, water is adhesive and elastic, and tends to aggregate in drops rather than spread out over a surface as a thin film. This phenomenon also causes water to stick to the sides of vertical structures despite gravity's downward pull. Water's high surface tension allows for the formation of water droplets and waves, allows plants to move water (and dissolved nutrients) from their roots to their leaves, and the movement of blood through tiny vessels in the bodies of some animals.

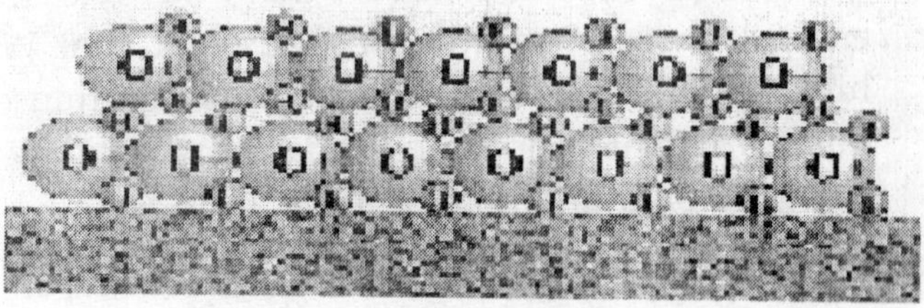

Figure 3: Water molecules are attracted to each other to create high surface tension.

— Water is the only substance on Earth that exists in all three *physical states of matter: solid, liquid,* and *gas.* Incorporated in the changes of state are massive amounts of heat exchange. This feature plays an important role in the redistribution of heat energy in the Earth's atmosphere. In terms of heat being transferred into the atmosphere, approximately 3/4's of this process is accomplished by the evaporation and condensation of water.

— The freezing of water causes it to expand. When water freezes it expands rapidly adding about 9 % by volume. Fresh water has a maximum density at around 4° Celsius (see Table 4). Water is the only substance on this planet where the maximum density of its mass does not occur when it becomes solidified.

Table 4: Density of water at various temperatures.

Temperature (degrees Celsius)	Density (grams per cubic centimeter)
0 (solid)	0.9150
0 (liquid)	0.9999
4	1.0000
20	0.9982
40	0.9922
60	0.9832
80	0.9718
100 (gas)	0.0006

Atmospheric Humidity

The term humidity describes the fact that the atmosphere can contain water vapor. The amount of humidity found in air varies because of a number of factors. Two important factors are evaporation and condensation. At the water/atmosphere interface over our planet's oceans large amounts of liquid water are evaporated into atmospheric water vapor. This process is mainly caused by absorption of solar radiation and the subsequent generation of heat at the ocean's surface. In our atmosphere, water vapor is converted back into liquid form when air masses lose heat energy and cool. This process is responsible for the development of most clouds and also produces the rain that falls to the Earth's surface.

Scientists have developed a number of different measures of atmospheric humidity. We are primarily interested in three of these measures: mixing ratio, saturation mixing ratio, and relative humidity. Mixing ratio is a measure that refers to the mass of a specific gas component relative to the mass of the remaining gaseous components for a sample of air.

When used to measure humidity mixing ratio would measure the mass of water vapor relative to the mass of all of the other gases. In meteorological measurements, mixing ratio is usually expressed in grams of water vapor per kilogram of dry air. Saturation mixing ratio refers to the mass of water vapor that can be held in a kilogram of dry air at saturation. Saturation can be generally defined as the condition where any addition of water vapor to a mass of air leads to the condensation of liquid water or the deposition of ice at a given temperature and pressure.

Table 5: Saturation mixing ratio (at 1000 mb).

Temperature Degrees Celsius	Vapor (g) per Kilogram of Dry Air
50	88.12
40	49.81
30	27.69
20	14.85
10	7.76

The data in Table 5 indicates that warmer air has a higher saturation mixing ratio than cooler air at a constant atmospheric pressure. It is important to note that this relationship between temperature and water vapor content in the air is not linear but exponential. In other words, for each 10° increase in temperature, saturation mixing ratio increases by a larger quantity.

The most commonly used measure of humidity is relative humidity. Relative humidity can be simply defined as the amount of water in the air relative to the saturation amount the air can hold at a given temperature multiplied by 100. Air with a relative humidity of 50 % contains a half of the water vapor it could hold at a particular temperature.

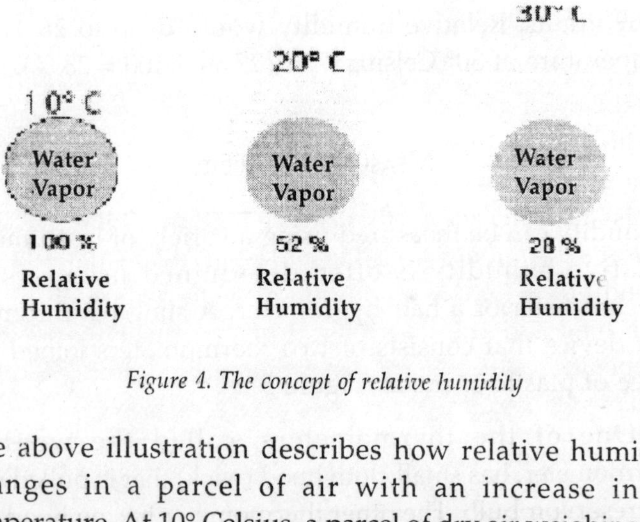

Figure 4. The concept of relative humidity

The above illustration describes how relative humidity changes in a parcel of air with an increase in air temperature. At 10° Celsius, a parcel of dry air weighing one kilogram can hold a maximum of 7.76 grams of water vapor (see Table 5). In this state, the parcel of air would be at saturation and its relative humidity would be 100 %. Increasing the temperature of this parcel, without adding or removing any water, would increase its ability to hold water vapor.

According to Table 5, a 10 degree Celsius rise in temperature would increase the saturation mixing ratio of this parcel of air to 14.85 grams. Since no water has been added or removed, the actual amount of water in the parcel would remain 7.76 grams. This quantity is known as the mixing ratio. Dividing the mixing ratio by the saturation mixing ratio and then multiplying this number by 100 determines the relative humidity of the parcel of air (7.76/ 14.85 x 100 = 52 %).

At a temperature of 20° Celsius, relative humidity would be 52 %. Raising the temperature of the parcel of air by another 10° Celsius would again lower its relative humidity. In this state, the actual mixing ratio would still be 7.76 grams, while the saturation mixing ratio would increase to 27.69 grams. Relative humidity would drop to 28 % at a temperature of 30° Celsius (7.76/27.69 × 100 = 28 %).

Measuring Humidity

Humidity can be measured using a variety of instruments. Relative humidity is often determined using a sling psychrometer or a hair hygrometer. A sling psychrometer is a device that consists of two thermometers joined to a piece of plastic or metal (Figure 5).

One of the thermometers, called the wet-bulb thermometer, has small cloth hood (wick) that is pulled over the reservoir bulb. The other thermometer has no hood and is called the dry-bulb thermometer. At one end of instrument is a rotating handle. To use the sling psychrometer, the wick is moistened with clean water and the device is twirled in the air using the handle. As the device is spun in the air, evaporation of the water from the wet-bulb thermometer occurs cooling it.

The amount of evaporation and cooling taking place is controlled by the dryness of the air. If the air is saturated, the wet-bulb and dry-bulb thermometers would have the same temperature because no evaporation can occur. After a few minutes of twirling, the temperatures of the wet-bulb and dry-bulb thermometers are determined, a value called the wet-bulb depression is calculated (dry-bulb minus wet-bulb temperature), and a psychrometric table is used to find the corresponding relative humidity from the dry-bulb temperature and wet-bulb depression.

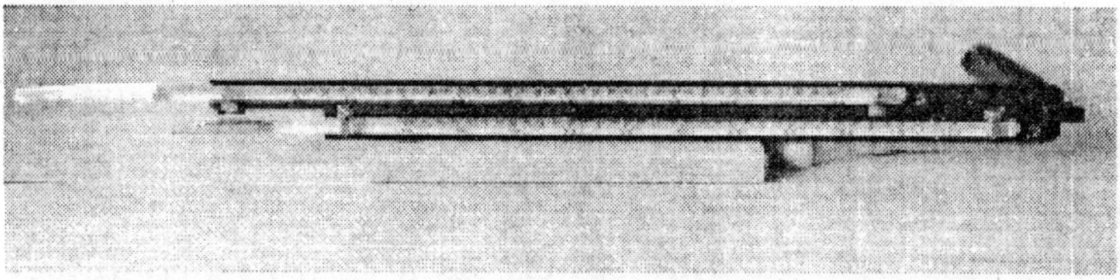

Figure 5: Sling psychrometer

Hair hygrometers work on the fact that hair changes its length when humidity varies. This device usually consists of a number of human or horse hairs connected to a mechanical lever system. When humidity increases the length of the hairs becomes longer. This change in length is then transmitted and magnified by the lever system into a measurement of relative humidity.

Humidity is also measured on a global scale using remotely placed satellites. These satellites are able to detect the concentration of water in the troposphere at altitudes between 4 and 12 kilometers. Satellites that can measure water vapor have sensors that are sensitive to infrared radiation. Water vapor specifically absorbs and re-radiates radiation in this spectral band. Satellite water vapor imagery plays an important role in monitoring climate conditions (like the formation of thunderstorms) and in the development of future weather forecasts.

Dew Point and Frost Point

Associated with relative humidity is dew point (if the dew point is below freezing, it is referred to as the frost point). Dew point is the temperature at which water vapor saturates from an air mass into liquid or solid usually forming rain, snow, frost, or dew. Dew point normally occurs when a mass of air has a relative humidity of 100 %. This happens in the atmosphere as a result of cooling through a number of different processes.

Condensation, Freezing, and Deposition

We have learned that water is available on the Earth in the following three forms: vapor; liquid; and solid. The process of water moving from one of these forms to another is called a phase or state change. In the atmosphere, three processes act to create water droplets or ice crystals. These three processes are:

— Condensation - water moving from a vapor to a liquid state.

— Freezing - water moving from a liquid to a solid state.

— Deposition - water moving from a vapor to a solid state.

For a phase change to occur heat energy must be added to or removed from water molecules. The formation of water droplets and ice crystals takes place when the water in the atmosphere is cooled. As air containing water vapor cools, the relative humidity of the air parcel increases until the dew or frost point is reached. At dew point (relative humidity = 100 %) water begins to condense into droplets. If 100 % relative humidity is reached below 0° Celsius deposition occurs and ice crystals form.

Formation of water droplets and ice crystals also requires a surface for condensation, freezing, or deposition. In the atmosphere, these surfaces are microscopic particles of dust, smoke, and salt commonly called condensation nuclei. Deposition nuclei, six sided particles, are needed for the formation of ice crystals.

If there is a deficiency of nuclei, super-saturation can result and condensation, freezing, or deposition can only occur with a relative humidity that is greater than 100 %

Cloud Formation Processes

Condensation or deposition of water above the Earth's surface creates clouds. In general, clouds develop in any air mass that becomes saturated (relative humidity becomes 100 %). Saturation can occur by way of atmospheric mechanisms that cause the temperature of an air mass to be cooled to its dew point or frost point. The following mechanisms or processes can achieve this outcome causing clouds to develop:

1. Orographic uplift occurs when air is forced to rise because of the physical presence of elevated land. As the parcel rises it cools as a result of adiabatic expansion at a rate of approximately 10° Celsius per 1000 meters until saturation. The development of

clouds and resulting heavy quantities of precipitation along the west coast of Canada are mainly due to this process.

2. Convectional lifting is associated with surface heating of the air at the ground surface. If enough heating occurs, the mass of air becomes warmer and lighter than the air in the surrounding environment, and just like a hot air balloon it begins to rise, expand, and cool. When sufficient cooling has taken place saturation occurs forming clouds. This process is active in the interior of continents and near the equator forming cumulus clouds and or cumulonimbus clouds (thunderstorms). The rain that is associated with the development of thunderstorm clouds is delivered in large amounts over short periods of time in extremely localized areas.

3. Convergence or frontal lifting takes place when two masses of air come together. In most cases, the two air masses have different temperature and moisture characteristics. Oneof the air masses is usually warm and moist, while the other is cold and dry. The leading edge of the latter air mass acts as an inclined wall or front causing the moist warm air to be lifted. Of course the lifting causes the warm moist air mass to cool due to expansion resulting in saturation. This cloud formation mechanism is common at the mid-latitudes where cyclones form along the polar front and near the equator where the trade winds meet at the intertropical convergence zone.

4.. Radiative cooling occurs when the sun is no longer supplying the ground and overlying air with energy derived from solar insolation (e.g., night). Instead, the surface of the Earth now begins to lose energy in the form of longwave radiation which causes the ground and air above it to cool. The clouds that result from this type of cooling take the form of surface fog.

Of course these causes of cloud development do not always act in a singular fashion. It is possible to get combinations of all four types, such as when convection and orographic uplift cause summer afternoon cloud development and showers in the mountains.

PRECIPITATION AND FOG

Precipitation

We can define precipitation as any aqueous deposit, in liquid or solid form, that develops in a saturated atmosphere (relative humidity equals 100 %) and falls to the ground generally from clouds. Most clouds, however, do not produce precipitation. In many clouds, water droplets and ice crystals are too small to overcome natural updrafts found in the atmosphere. As a result, the tiny water droplets and ice crystals remain suspended in the atmosphere as clouds.

Water droplets and ice crystals can only fall to the Earth's surface if they grow to a size that can overcome updrafts. Conditions for growth can develop in clouds via two different processes.

In clouds with temperatures above freezing, turbulent atmospheric mixing can cause droplets to grow through collision and coalescence. One initial condition, however, must be met for this process to begin: droplet size in the cloud must be variable. This initial condition allows larger and heavier droplets to collide and coalesce with lighter smaller droplets during downdraft periods. If enough atmospheric mixing occurs the larger droplets can expand by up to 250 times and can become heavy enough to fall to the Earth's surface.

The other mechanism of precipitation development involves clouds whose temperature is below freezing. In these clouds, large ice crystals grow due to the differences in vapor pressure between ice crystals and supercooled water droplets. Vapor pressure differences between ice and supercooled water causes a net migration of water vapor from water droplets to ice crystals. The ice crystal then absorbs the water vapor, depositing it on their surface. At the same time, the loss of vapor from the water droplets causes them to shrink in size.

A necessary initial requirement for this process is the presence of both condensation nuclei and deposition nuclei. While deposition nuclei form ice crystals at temperatures just below zero degrees Celsius, condensation nuclei can remain liquid (supercooled) to temperatures as low as - 40° Celsius depending on size. Because of this phenomenon, cold clouds can contain both ice crystals and supercooled water droplets. The relative proportion of these two types of particles determines whether snow crystals grow to a size to overcome atmospheric updrafts.

The following list describes the various types of precipitation that can form in the atmosphere:

Rain is any liquid deposit that falls from the atmosphere to the surface and has a diameter greater than 0.5 millimeters.

Freezing rain occurs when water droplets hit a cold surface and freeze. For this to occur a surface temperature inversion is required. In such an inversion, the surface must have a temperature below freezing, while the temperature of the atmosphere where the precipitation forms is above freezing. Surface inversions may develop from a variety of causes, but typically, they occur near the leading edge of cold air from the North as it pushes southward in the Northern Hemisphere (in the Southern Hemisphere surface inversions occur near the leading edge of cold air from the South as it pushes northward).

Ice pellets or sleet are transparent or translucent bits of frozen water with a diameter less than 5 millimeters. To form, these pellets require an environment where raindrops develop in an atmosphere where the temperature is above freezing and then fall into a lower layer of air with temperatures below freezing. In the lower layer of cold air the raindrops freeze into small ice pellets. Normally, to provide temperatures above freezing in the atmosphere above a colder atmospheric layer an upper air temperature inversion is required.

Snow can develop only from water vapor that deposits directly as a solid on a deposition nuclei at temperatures below freezing by passing the liquid state. A snowflake forms first as a very tiny crystal developing on a six-sided hexagonal deposition nuclei. The ice crystal then grows fastest at the six points as these areas are more directly exposed to the atmosphere's water vapor. Snow is most common in winter just North of the center of mid-latitude cyclones in the Northern Hemisphere (south of a mid-latitude in the Southern Hemisphere).

As the warm moist air travels around the center of lowest pressure, it overrides colder air located North of the low and is cooled to its saturation temperature, producing rainfall and snow. Snow generally occurs with East winds, since the winds at locations north of a mid-latitude cyclone are from the East. Much of the ground surface of North America can be covered with snow during the winter months (Figure 6).

Figure 6: Close-up photograph of a snowflake.

Snow pellets are white, spherical grains of ice 2 to 5 millimeters in diameter. They can be distinguished from packed snowflakes since snow pellets are firm enough to bounce when they hit the ground. Snow pellets develop as supercooled droplets freeze on ice crystals. They may fall for a brief period as the precipitation changes from ice pellets to snow.

Hail is a destructive form of precipitation that is 5 to 190 millimeters in diameter. The large downdrafts in mature thunderstorm clouds provide the mechanism for hail formation. Hailstones normally have concentric shells of ice alternating between those with a milky appearance and those that are clear. The milky white shells, containing bubbles and partially melted snowflakes, correspond to a period of rapid freezing, while the clear shells develop as the liquid water freezes much more slowly.

Fog

If the air near the ground is cooled sufficiently, it becomes saturated and fog can develop. By definition, fog exists if the atmospheric visibility near the Earth's surface is reduced to 1 kilometer or less. Fogs composed primarily of water droplets can be classified according to the process that causes the air to cool and saturate:

— Radiation fog or ground fog, is generated as the Earth's surface cools by radiation loss at night. This type of fog is normally quite shallow.

— Upslope fog is generated by airflow over topographic barriers. As the air is forced to rise upward where the atmospheric pressure is less, it is cooled by expansion and produces fog on the windward slopes of hills or mountains.

— Advection fog may be generated by winds that contrast in temperature with the Earth's surface. Warm air advection can produce fog through contact cooling with a cold surface.

— Evaporation fog occurs when you get cold air advection over warm water or warm, moist land surfaces resulting in fog formation as water evaporates into the cold air. This type of fog is sometimes called steam fog or sea smoke.

— Frontal fog is produced as weather fronts, especially warm fronts, pass through an area. Precipitation falling into the colder air ahead of the warm front may evaporate enough water to cause the formation of small droplets as fog near the ground.

Global Distribution of Precipitation

The Global Precipitation Climatology Project (GPCP) was established by the World Climate Research Program (WCRP) in 1986 with the goal of providing monthly mean precipitation data on a 2.5 × 2.5 degree latitude-longitude grid for the period 1986-2000. The GPCP will accomplish this by combining infrared and microwave satellite estimates of precipitation with rain gauge data from more than 30,000 stations. Infrared precipitation measurements are obtained from GOES, GMS and Meteosat (European Community) geostationary satellites and National Oceanic and Atmospheric Administration (NOAA) operational polar orbiting satellites. Microwave estimates are obtained from the U.S. Defense Meteorological Satellite Program (DMSP) satellites using the Special Sensor Microwave Imager (SSM/I). Together these data sets will be used to validate general circulation and climate models, study the global hydrological cycle and diagnose the variability of the global climate system.

The average annual precipitation of the world is estimated to be 1050 millimeters per year or 2.9 millimeters per day. However, actual values vary from a minimum of 0 millimeters per day or to a maximum of 10 millimeters per day depending on location. The reasons for these patterns are as follows:

The deserts in the subtropical regions occur because these areas do not contain any mechanism for lifting air masses. In fact, these areas are dominated by subsiding air that results from global circulation patterns.

Continental areas tend to be dry because of their distance from moisture sources. Polar areas are dry because cold air cannot hold as much moisture as warm air. Areas near the equator achieve high rainfall amounts because constant solar heating encourages convection, and global circulation patterns cause northern and southern air masses to converge here causing frontal lifting.

Mid-latitudes experience cyclonic activity and frontal lifting when polar and subtropical air masses meet at the polar front. Further, the air masses in this region generally move from West to East, causing levels of precipitation to decrease East of source regions.

Mountain ranges near water sources can receive high rainfalls because of orographic uplift, if and only if the prevailing winds are in their favor. This can also result in a sharp reduction in rainfall in regions adjacent or on the leeward slopes of these areas. This phenomenon is commonly know as the rainshadow effect.

Acid Precipitation

Acidic pollutants can be deposited from the atmosphere to the Earth's surface in wet and dry forms. The common term to describe this process is acid deposition. The term acid precipitation is used to specifically describe wet forms of acid pollution that can be found in rain, sleet, snow, fog, and cloud vapor. An acid can be defined as any substance that when dissolved in water dissociates to yield corrosive hydrogen ions. The acidity of substances dissolved in water is commonly measured in terms of pH (defined as the negative logarithm of the concentration of hydrogen ions). According to this measurement scale solutions with pHs less than 7 are described as being acidic, while a pH greater than 7.0 is considered alkaline. Precipitation normally has a pH between 5.0 to 5.6 because of natural atmospheric reactions involving carbon dioxide. For comparison, distilled water, pure of any other stubstances, would have a pH of 7.0. Precipitation is considered to be acidic when its pH falls below 5.6 (which is 25 times more acidic than pure distilled water). Some sites in eastern North America have precipitation events with pHs as low as 2.3 or about 1000 times more acidic than natural.

Acid deposition is not a recent phenomenon. In the 17th century, scientists noted the ill effects that industry and acidic pollution was having on vegetation and people. However, the term acid rain was first used two centuries later when Angus Smith published a book called 'Acid Rain' in 1872. In the 1960s, the problems associated with acid deposition became an international problem when fishermen noticed declines in fish numbers and diversity in many lakes throughout North America and Europe.

Acid Deposition Formation

Acid deposition can form as a result of two processes. In some cases, hydrochloric acid can be expelled directly into the atmosphere. More commonly it is due to secondary pollutants that form from the oxidation of nitrogen oxides (NO_x) or sulfur dioxide (SO_2) gases that are released into the atmosphere. Reactions at the Earth's surface or within the atmosphere can convert these pollutants into nitric acid or sulfuric acid. The process of altering these gases into their acid counterparts can take several days, and during this time these pollutants can be transported hundreds of kilometers from their original source.

Acid precipitation formation can also take place at the Earth's surface when nitrogen oxides and sulfur dioxide settle on the landscape and interact with dew or frost. Emissions of sulfur dioxide are responsible for 60-70 % of the acid deposition that occurs globally. More than 90 % of the sulfur in the atmosphere is of human origin. The main sources of sulfur include:

- Coal burning - coal typically contains 2-3 % sulfur so when it is burned sulfur dioxide is liberated.
- The smelting of metal sulfide ores to obtain the pure metals. Metals such as zinc, nickel, and copper are all commonly obtained in this manner.
- Volcanic eruptions - although this is not a widespread problem, a volcanic eruption can add a lot of sulfur to the atmosphere in a regional area.

Organic decay

Ocean spray: After being released into the atmosphere, sulfur dioxide can either be deposited on the Earth's surface in the form of dry deposition or it can undergo the following reactions to produce acids that are incorporated into the products of wet deposition:

$$SO_2 + H_2O \text{ »»» } H_2SO_3$$

$$H_2SO_3 + 1/2O_2 \text{ »»» } H_2SO_4$$

Nitrogen oxides (NO_x) and sulfur dioxide (SO_2) released into the atmosphere from a variety of sources call fall to the ground simply as dry deposition. This dry deposition can then be converted into acids when these deposited chemicals meet water. Most wet acid deposition forms when nitrogen oxides (NOx) and sulfur dioxide (SO_2) are converted to nitric acid (HNO_3) and sulfuric acid (H_2SO_4) through oxidation and dissolution. Wet deposition can also form when ammonia gas (NH_3) from natural sources is converted into ammonium (NH_4).

Some 95 % of the elevated levels of nitrogen oxides in the atmosphere are the result of human activities. The remaining 5 % comes from several natural processes. The major sources of nitrogen oxides include:

— Combustion of oil, coal, and gas.

— Bacterial action in soil.

— Forest fires.

— Volcanic action.

— Lightning.

Acids of nitrogen form as a result of the following atmospheric chemical reactions:

$$NO + 1/2O_2 \text{ »»» } NO_2$$

$$2NO_2 + H_2O \text{ »»» } HNO_2 + HNO_3$$

$$NO_2 + OH \text{ »»» } HNO_3$$

Finally, the concentrations of both nitrogen oxides and sulfur dioxides are much lower than atmospheric carbon dioxide which is mainly responsible for making natural rainwater slightly acidic. However, these gases are much more soluble than carbon dioxide and therefore have a much greater effect on the pH of the precipitation.

Effects of Acid Deposition

Acid deposition influences the environment in several different ways. In aquatic systems, acid deposition can effect these ecosystems by lowering their pH. However, not all aquatic systems are effected equally. Streams, ponds, or lakes that exist on bedrock or sediments rich in calcium and/or magnesium are naturally buffered from the effects of acid deposition. Aquatic systems on neutral or acidic bedrock are normally very sensitive to acid deposition because they lack basic compounds that buffer acidification.

In Canada, many of the water bodies found on the granitic Canadian Shield fall in this group. One of the most obvious effects of aquatic acidification is the decline in fish numbers. Originally, it was believed that the fish died because of the increasing acidity of the water. However, in the 1970s scientists discovered that acidified lakes also contained high concentrations of toxic heavy metals like mercury, aluminum, and cadmium.

The source of these heavy metals was the soil and bedrock surrounding the water body. Normally, these chemicals are found locked in clay particles, minerals, and rocks. However, the acidification of terrestrial soils and bedrock can cause these metals to become soluble. Once soluble, these toxic metals are easily leached by infiltrating water into aquatic systems where they accumulate to toxic levels.

The acidity of the leachate entering lake is controlled by the chemical composition of the effected lake's surrounding soil and bedrock. If the soil and bedrock is rich in limestone the acidity of the infiltrate can be reduced by the buffering action of calcium and magnesium compounds. Toxic aluminum (and some other toxic heavy metals) can leach into the lake if the soil and bedrock is rich in aluminum-rich silicate minerals.

In the middle latitudes, many acidified aquatic systems experience a phenomenon known as acid shock. During the winter the acidic deposits can buildup in the snowpack. With the arrival of spring, snowpack begins to melt quickly and the acids are released over a short period of time at concentrations 5 to 10 times more acidic than rainfall. Most adult fish can survive this shock. However, the eggs and small fry of many spring spawning species are extremely sensitive to this acidification.

The severity of the impact of acid deposition on vegetation is greatly dependent on the type of soil the plants grow in. Similar to surface water acidification, many soils have a natural buffering capacity and are able to neutralize acid inputs. In general, soils that have a lot of lime are better at neutralizing acids than those that are made up of siliceous sand or weathered acidic bedrock. In less buffered soils, vegetation is effected by acid deposition because:

— Increasing acidity results in the leaching of several important plant nutrients, including calcium, potassium, and magnesium. Reductions in the availability of these nutrients cause a decline in plant growth rates.

— The heavy metal aluminum becomes more mobile in acidified soils. Aluminum can damage roots and interfere with plant uptake of other nutrients such as magnesium and potassium.

— Reductions in soil pH can cause germination of seeds and the growth of young seedlings to be inhibited.

— Many important soil organisms cannot survive is soils below a pH of about 6.0. The death of these organisms can inhibit decomposition and nutrient cycling.

— High concentrations of nitric acid can increase the availability of nitrogen and reduce the availability of other nutrients necessary for plant growth. As a result, the plants become over-fertilized by nitrogen (a condition known as nitrogen saturation).

— Acid precipitation can cause direct damage to the foliage on plants especially when the precipitation is in the form of fog or cloud water which is up to ten times more acidic than rainfall.

— Dry deposition of SO_2 and NOx has been found to affect the ability of leaves to retain water when they are under water stress.

— Acidic deposition can leach nutrients from the plant tissues weakening their structure.

— The combination of these effects can lead to plants that have reduced growth rates, flowering ability and yields. It also makes plants more vulnerable to diseases, insects, droughts and frosts.

The effects of acidic deposition on humans can be divided into three main categories. Acid deposition can influence human health through the following methods:

Toxic metals, such as mercury and aluminum, can be released into the environment through the acidification of soils. The toxic metals can then end up in the drinking water, crops, and fish, and are then ingested by humans through consumption. If ingested in great quantities, these metals can have toxic effects on human health. One metal, aluminum, is believed to be related to the occurrence of Alzheimer's disease.

Increased concentrations of sulfur dioxide and oxides of nitrogen have been correlated to increased hospital admissions for respiratory illness.

Research on children from communities that receive a high amount of acidic pollution show increased frequencies of chest colds, allergies, and coughs.

Acid deposition also influences the economic livelihoods of some people. Many lakes and streams on the eastern coast of North America are so acidic that the fish decline significantly in numbers. The reduced fish numbers then influence commercial fishermen and industries that rely on sport fishing tourism. Forestry and agriculture are effected by the damage caused to vegetation. In some areas of eastern North America and Europe, large die-backs of trees have occurred.

Finally, acid deposition effects a number inanimate features of human construction. Buildings and head stones that are constructed from limestone are easily attacked by acids, as are structures that are constructed of iron or steel. Paint on cars can react with acid deposition causing fading. Many of the churches and cathedrals in Europe are under attack from the effects of acidic deposition.

Evaporation and Transpiration

Water is removed from the surface of the Earth to the atmosphere by two distinct mechanisms: evaporation and transpiration.

Evaporation can be defined as the process where liquid water is transformed into a gaseous state. Evaporation can only occur when water is available. It also requires that the humidity of the atmosphere be less than the evaporating surface (at 100% relative humidity there is no more evaporation). The evaporation process requires large amounts of energy. For example, the evaporation of one gram of water requires 600 calories of heat energy.

Transpiration is the process of water loss from plants through stomata. Stomata are small openings found on the underside of leaves that are connected to vascular plant tissues. In most plants, transpiration is a passive process largely controlled by the humidity of the atmospheric and the moisture content of the soil. Of the transpired water passing through a plant only 1 % is used in the growth process.

Transpiration also transports nutrients from the soil into the roots and carries them to the various cells of the plant and is used to keep tissues from becoming overheated. Some dry environment plants do have the ability to open and close their stomata. This adaptation is necessary to limit the loss of water from plant tissues. Without this adaptation these plants would not be able to survive under conditions of severe drought.

It is often difficult to distinguish between evaporation and transpiration. So we use a composite term evapotranspiration. The rate of evapotranspiration at any instant from the Earth's surface is controlled by four factors:

— *Energy availability*: The more energy available the greater the rate of evapotranspiration. It takes about 600 calories of heat energy to change 1 gram of liquid water into a gas.

— *The humidity gradient away from the surface*: The rate and quantity of water vapor entering into the atmosphere both become higher in drier air.

— *The wind speed immediately above the surface*: Many of us have observed that our gardens need more watering on windy days compared to calm days when temperatures are similar. This fact occurs because wind increases the potential for evapotranspiration. The process of evapotranspiration moves water vapor from ground or water surfaces to an adjacent shallow layer that is only a few centimeters thick. When this layer becomes saturated evapotranspiration stops. However, wind can remove this layer replacing it with drier air which increases the potential for evapotranspiration.

— *Water availability*: Evapotranspiration cannot occur if water is not available.

On a global scale, most of the evapotranspiration of water on the Earth's surface occurs in the subtropical oceans. In these areas, high quantities of solar radiation provide the energy required to convert liquid water into a gas. Evapotranspiration generally exceeds precipitation on middle and high latitude landmass areas during the summer season. Once again, the greater availability of solar radiation during this time enhances the evapotranspiration process.

Actual and Potential Evapotranspiration

Often, scientists distinguish between two different aspects of evapotranspiration: potential evapotranspiration and actual evapotranspiration.

Potential evapotranspiration or PE is a measure of the ability of the atmosphere to remove water from the surface through the processes of evaporation and transpiration assuming no control on water supply. Actual evapotranspiration or AE is the quantity of water that is actually removed from a surface due to the processes of evaporation and transpiration.

Scientists consider these two types of evapotranspiration for the practical purpose of water resource management.

Around the world humans are involved in the production of a variety of plant crops. Many of these crops grow in environments that are naturally short of water. As a result, irrigation is used to supplement the crop's water needs. Managers of these crops can determine how much supplemental water is needed to achieve maximum productivity by estimating potential and actual evapotranspiration. Estimates of these values are then used in the following equation:

crop water need = potential evapotranspiration - actual evapotranspiration

The following factors are extremely important in estimating potential evapotranspiration:

— Potential evapotranspiration requires energy for the evaporation process. The major source of this energy is from the sun. The amount of energy received from the sun accounts for 80 % of the variation in potential evapotranspiration.

— Wind is the second most important factor influencing potential evapotranspiration. Wind enables water molecules to be removed from the ground surface by a process known as eddy diffusion.

— The rate of evapotranspiration is associated to the gradient of vapor pressure between the ground surface and the layer of atmosphere receiving the evaporated water.

Interception, Stemflow, Canopy Drip, and Throughfall

Vegetation often modifies the intensity and distribution of precipitation falling on and through its leaves and woody structures. The most obvious effect plants have on falling precipitation is interception. Interception can be technically defined as the capture of precipitation by the plant canopy and its subsequent return to the atmosphere through evaporation or sublimation. The amount of precipitation intercepted by plants varies with leaf type, canopy architecture, wind speed, available radiation, temperature, and the humidity of the atmosphere.

Precipitation that is not intercepted can be influenced by the following processes:

— Stemflow - is the process that directs precipitation down plant branches and stems. The redirection of water by this process causes the ground area around the plant's stem to receive additional moisture. The amount of stemflow is determined by leaf shape and stem and branch architecture. In general, deciduous trees have more stemflow than coniferous vegetation.

— Canopy drip - some plants have an architecture that directs rainfall or snowfall along the edge of the plant canopy. This is especially true of coniferous vegetation. On the ground, canopy drip creates areas with higher moisture content that are located in a narrow band at the edge of the plant canopy.

— Throughfall - describes the process of precipitation passing through the plant canopy. This process is controlled by factors like: plant leaf and stem density, type of the precipitation, intensity of the precipitation, and duration of the precipitation event. The amount of precipitation passing through varies greatly with vegetation type.

Infiltration and Soil Water Storage

Infiltration refers to the movement of water into the soil layer. The rate of this movement is called the infiltration rate. If rainfall intensity is greater than the infiltration rate, water will accumulate on the surface and runoff will begin.

Movement of water into the soil is controlled by gravity, capillary action, and soil porosity. Of these factors soil porosity is most important. A soil's porosity is controlled by its texture, structure, and organic content. Coarse textured soils have larger pores and fissures than fine-grained soils and therefore allow for more water flow. Pores and fissures found in soils can be made larger through a number of factors that enhance internal soil structure.

For example, the burrowing of worms and other organisms and penetration of plant roots can increase the size and number of macro and micro-channels within the soil. The amount of decayed organic matter found at the soil surface can also enhance infiltration. Organic matter is generally more porous than mineral soil particles and can hold much greater quantities of water.

The rate of infiltration normally declines rapidly during the early part of a rainstorm event and reaches a constant value after several hours of rainfall. A number of factors are responsible for this phenomena, including:

(1) The filling of fine soil pores with water reduces capillary forces.

(2) As the soil moistens, clay particles swell and reduce the size of pores.

(3) Raindrop impact breaks up soil clumps, splashing fine particles into pores.

Soil Water Storage

Within the soil system, the storage of water is influenced by several different forces. The strongest force is the molecular force of elements and compounds found on the surface of soil minerals. The water retained by this force is called hygroscopic water and it consists of the water held within 0.0002 millimeters of the surface of soil particles.

The maximum limit of this water around a soil particle is known as the hygroscopic coefficient. Hygroscopic water is essentially non-mobile and can only be removed from the soil through heating. Matric force holds soil water from 0.0002 to 0.06 millimeters from the surface of soil particles. This force is due to two processes: soil particle surface molecular attraction (adhesion and absorption) to water and the cohesion that water molecules have to each other. This force declines in strength with distance from the soil particle.

The force becomes nonexistent past 0.06 millimeters. Capillary action moves this water from areas where the matric force is low to areas where it is high. Because this water is primarily moved by capillary action, scientists commonly refer to it as capillary water. Plants can use most of this water by way of capillary action until the soil wilting point is reached. Water in excess of capillary and hygroscopic water is called gravitational water. Gravitational water is found beyond 0.06 millimeters from the surface of soil particles and it moves freely under the effect of gravity. When gravitational water has drained away the amount of water that remains is called the soil's field capacity.

Throughflow and Groundwater Storage

Throughflow is the sporadic horizontal flow of water within the soil layer. It normally takes place when the soil is completely saturated with water. This water then flows underground until it reaches a river, lake, or ocean. Rates of water movement via throughflow are usually low. Rates of maximum flow occur on steep slopes and in pervious sediments. The lowest rates of flow occur in soils composed of heavy clays. Rates of throughflow in these sediments can be less than 1 millimeter per day.

Precipitation that succeeds in moving from the soil layer down into the underlying bedrock will at some point reach an area of permanent saturation that is known as the groundwater zone. The top of this zone is called the water table. Approximately 22 % of the fresh water found at the Earth's terrestrial surface is stored as groundwater. Groundwater tends to flow by way of gravity to the point of lowest elevation. Often groundwater flow discharges into a surface body of water like a river channel, lake, or ocean.

Typical groundwater flow velocities lie in the range of 250 to 0.001 meters per day. Highest groundwater flow velocities are commonly found in sedimentary deposits (like gravel, conglomerate, or sandstone) because of their very high permeability. The least permeable ground type occurs in dense igneous rock materials like granite. Rock formations that store groundwater water are known as aquifers. Rock formations that cannot store groundwater are called aquicludes.

Groundwater occurs in two main forms. Unconfined groundwater occurs when the flow of subterranean water is not confined by the presence of relatively impermeable layers. The presence of an impermeable layer beneath this type of groundwater can cause the formation of a perched water table. These features are elevated some distance above the surface's main water table. Springs that flow from underground to the Earth's surface are often formed when a perched water table intersects the surface.

In some cases, groundwater can become confined between two impermeable layers. This type of enclosed water is sometimes called artesian. If conditions are right, a confined aquifer can produce a pressurized ground to surface flow of water known as an artesian well. In an artesian well, water flows against gravity to the earth's surface because of hydrostatic pressure. Hydrostatic pressure is created from the fact that most of the aquifer's water resides at an elevation greater than the well opening. The overlying weight of this water creates the hydrostatic pressure.

Surface Runoff

If the amount of water falling on the ground is greater than the infiltration rate of the surface, runoff or overland flow will occur. Runoff specifically refers to the water leaving an area of drainage and flowing across the land surface to points of lower elevation. It is not the water flowing beneath the surface of the ground. This type of water flow is called throughflow. Runoff involves the following events:

— Rainfall intensity exceeds the soil's infiltration rate.

— A thin water layer forms that begins to move because of the influence of slope and gravity.

— Flowing water accumulates in depressions.

— Depressions overflow and form small rills.

— Rills merge to form larger streams and rivers.

— Streams and rivers then flow into lakes or oceans.

On a global scale, runoff occurs because of the imbalance between evaporation and precipitation over the Earth's land and ocean surfaces. Oceans make up 71 % of the Earth's surface and the solar radiation received here powers the global evaporation process. In fact, 86 % of the Earth's evaporation occurs over the oceans, while only 14 % occurs over land. Of the total amount of water evaporated into the atmosphere, precipitation returns only 79 % to the oceans, and 21 % to the land. Surface runoff sends 7 % of the land based precipitation back to the ocean to balance the processes of evaporation and precipitation.

The distribution of runoff per continent shows some interesting patterns (see Table 6). Areas having the most runoff are those with high rates of precipitation and low rates of evaporation.

Table 6: Continental runoff values

Continent	Runoff Per Unit Area (mm per yr.)
Europe	300
Asia	286
Africa	139
North and Central America	265
South America	445
Australia, New Zealand and New Guinea	218
Antarctica and Greenland	164

Streamflow and Stream Discharge

The term streamflow describes the process of water flowing in the organized channels of a stream or river. Stream discharge represents the volume of water passing through a river channel during a certain period of time. Stream discharge can be expressed mathematically with the following equation:

$$Q = W \times D \times V$$

where,

Q equals stream discharge usually measured in cubic meters per second, W equals channel width, D equals channel depth, and V equals velocity of flowing water.

Because of streamflow's potential hazard to humans many streams are gauged by mechanical recorders. These instruments record the stream's discharge on a hydrograph. The graph (Figure 7) below illustrates a typical hydrograph and its measurement of discharge over time.

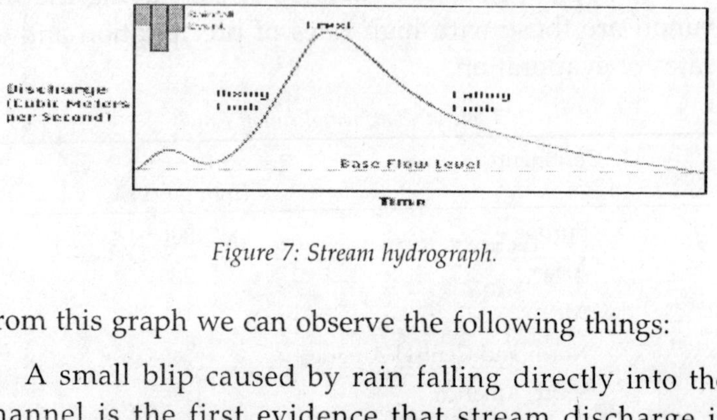

Figure 7: Stream hydrograph.

From this graph we can observe the following things:

A small blip caused by rain falling directly into the channel is the first evidence that stream discharge is changing because of the rainfall.

A significant time interval occurs between the start of rain and the beginning of the main rise in discharge on the hydrograph. This lag occurs because of the time required for the precipitation that falls in the stream's basin to eventually reach the recording station. Usually, the larger the basin the greater the the time lag.

The rapid movement of surface runoff into the stream's channels and subsequent flow causes the discharge to rise quickly.

The falling limb of the hydrograph tends to be less steep that the rise. This flow represents the water added from distant tributaries and from throughflow that occurs in surface soils and sediments.

After some time the hydrograph settles at a constant level known as base flow stage. Most of the base flow comes from groundwater flow which moves water into the stream channel very slowly.

Not all hydrographs are the same. Actually, the shape and magnitude of the hydrograph is controlled by two sets of factors:

— Permanent Factors - slope of basin, soil structure, vegetation, channel density, etc.

— Transient Factors - are those factors associated with precipitation input - size of storm, intensity, duration of rainfall, etc.

OCEANS

Oceans cover approximately 71 % or 360 million square kilometers of the Earth's surface. On average, the depth of the world's oceans is about 3.9 kilometers. Maximum depths, however, can exceed 11 kilometers! The oceans contain 97 % of our planet's free water. The other 3 % is found in atmosphere, on the Earth's terrestrial surface, or in the Earth's lithosphere in various forms and stores.

The distribution of ocean basins and continents is unevenly arranged over the Earth's surface. In the Northern

Hemisphere, the ratio of land to ocean is about 1 to 1.5. The ratio of land to ocean in the Southern Hemisphere is 1 to 4. The greater abundance of water in the Southern Hemisphere has some interesting effects on the environment of this area.

For example, climate tends to be more moderate in the Southern Hemisphere because of the ocean's ability to release large amounts of stored heat energy. Humans have divided and named the interconnected oceans of the world into three groups: the Atlantic (including the Arctic Sea), the Indian, and the Pacific.

The Pacific is the largest ocean basin. It has an average depth of 4.3 kilometers and has few shallow marginal seas, but many islands. Only a few rivers discharge into this ocean basin. This lack of rivers is demonstrated by fact that the surface area of the Pacific is about 1000 percent greater than the land area that drains into it.

The Atlantic is a relatively narrow body of water that twists between nearly parallel continental masses. The Atlantic Ocean contains the majority of the Earth's shallow seas, but relatively few islands. Some of the shallow seas found in this basin include: the Caribbean, Mediterranean, Baltic, Arctic Seas, and the Gulf of Mexico. Many streams and rivers discharge into the Atlantic Ocean. This basin also drains some of the world's largest rivers including the Amazon, Mississippi, St. Lawrence, and Congo. The surface area of the Atlantic Ocean is about 1.6 times greater than the terrestrial area discharging into to it. As a result, the Atlantic Ocean receives more fresh water from continental runoff than any other ocean basin.

The Indian Ocean is the smallest of the three major ocean basins. It is bordered by the landmasses of Africa and

Asia. This basin has few islands and limited shallow seas. The surface area of the Indian ocean is approximately 400 percent larger than the area draining into it. Because of its close proximity to the equator this basin has the warmest surface ocean temperatures.

Physical and Chemical Characteristics of Seawater

Seawater is a mixture of various salts and water. Most of the water in the ocean basins is believed to originate from the condensation of water found in the early atmosphere as the Earth cooled after its formation. This water was released from the lithosphere as the Earth's crust solidified. Additional water has also been added to the oceans over geologic time from periodic volcanic action. Some scientists have recently speculated that comets entering the Earth's atmosphere may be another important source of water for the oceans.

Most of the dissolved chemical constituents or salts found in seawater have a continental origin. It seems that these chemicals were released from continental rocks through weathering and then carried to the oceans by stream runoff. Over time, the concentration of these chemicals increased until an equilibrium was met. This equilibrium occurred when the ocean's water could not dissolve any more material in solution. Similarities between fossilized sea life and organisms living today indicate that the composition of seawater stopped changing drastically about 600 million years ago.

Only six elements and compounds comprise about 99 % of sea salts: chlorine (Cl^-), sodium (Na^+), sulfur ($SO4^{-2}$), magnesium (Mg^{+2}), calcium (Ca^{+2}), and potassium (K^+). The relative abundance of the major salts in seawater are constant regardless of the ocean. Only the amount of water in the mixture varies because of differences between ocean basins because of regional differences in freshwater loss (evaporation) and gain (runoff and precipitation).

The chlorine ion makes up 55 % of the salt in seawater. Calculations of seawater salinity are made of the parts per 1000 of the chlorine ion present in one kilogram of seawater. Typically, seawater has a salinity of 35 parts per thousand.

Water is one of the few substances existing on the Earth's surface in all three forms of matter. At zero degrees Celsius liquid water turns into ice and has a density of approximately 917 kilograms per cubic meter. Liquid water at the same temperature has a density of nearly 1,000 kilograms per cubic meter. The density of seawater generally increases with decreasing temperature, increasing salinity, and increasing depth in the ocean. The density of seawater at the surface of the ocean varies from 1,020 to 1,029 kilograms per cubic meter. Highest densities are achieved with depth because of the overlying weight of water. In the deepest parts of the oceans, seawater densities can be as high as 1,050 kilograms per cubic meter.

Seawater freezes at a temperature that is slightly colder than fresh water (0.0° Celsius). The freezing temperature of seawater also varies with the concentration of salts. More salt the lower the initial freezing temperature. At a salinity of 35 parts per thousand, seawater freezes at a temperature of -1.9° Celsius.

Sea ice normally contains considerably less salt than seawater. Most of the salts found in liquid seawater are forced out it when freezing occurs. The reason for the exclusion is because the molecules of the various salts do not fit well in the highly orderly molecular structure of frozen water. Because of the density difference between ice and seawater, ice floats on the surface of the ocean.

Seawater also contains small amounts of dissolved gases. Many of these gases are added to seawater from the atmosphere through the constant stirring of the sea surface by wind and waves. The concentration of gases that can be dissolved into seawater from the atmosphere is determined by temperature and salinity of the water. Increasing the temperature or salinity reduces the amount of gas that ocean water can dissolve. Some of the important atmospheric gases found in seawater include: nitrogen, oxygen, carbon dioxide (in the form of bicarbonate HCO_3), argon, helium, and neon. Compared to the other atmospheric gases, the amount of carbon dioxide dissolved in saturated seawater is unusually large.

Some gases found within seawater are also involved in oceanic organic and inorganic processes that are indirectly related to the atmosphere. For example, oxygen and carbon dioxide may be temporally generated or depleted by such processes to varying concentrations at specific locations within the ocean.

Surface and Subsurface Ocean Currents

An ocean current can be defined as a horizontal movement of seawater at the ocean's surface. Ocean currents are driven by the circulation of wind above surface waters. Frictional stress at the interface between the ocean and the wind causes the water to move in the direction of the wind. Large ocean currents are a response of the atmosphere and ocean to the flow of energy from the tropics to polar regions. In some cases, currents are transient features and affect only a small area. Other ocean currents are essentially permanent and extend over large horizontal distances.

On a global scale, large ocean currents are constrained by the continental masses found bordering the three oceanic basins. Continental borders cause these currents to develop an almost closed circular pattern called a gyre. Each ocean basin has a large gyre located at approximately 30° North and South latitude in the subtropical regions. The currents in these gyres are driven by the atmospheric flow produced by the subtropical high pressure systems. Smaller gyres occur in the North Atlantic and Pacific Oceans centered at 50° North. Currents in these systems are propelled by the circulation produced by polar low pressure centers. In the Southern Hemisphere, these gyre systems do not develop because of the lack of constraining land masses.

A typical gyre displays four types of joined currents: two east-west aligned currents found respectively at the top and bottom ends of the gyre; and two boundary currents oriented north-south and flowing parallel to the continental margins. Direction of flow within these currents is determined by the direction of the macro-scale wind circulation. Boundary currents play a role in redistributing global heat latitudinally.

Surface Currents of the Subtropical Gyres

On either side of the equator, in all ocean basins, there are two west flowing currents: the North and South Equatorial These currents flow between 3 and 6 kilometers per day and usually extend 100 to 200 meters in depth below the ocean surface. The Equatorial Counter Current, which flows towards the east, is a partial return of water carried westward by the North and South Equatorial currents. In El Nino years, this current intensifies in the Pacific Ocean.

Flowing from the equator to high latitudes are the western boundary currents. These warm water currents have specific names associated with their location: North Atlantic - Gulf Stream; North Pacific - Kuroshio; South Atlantic - Brazil; South Pacific - East Australia; and Indian Ocean - Agulhas. All of these currents are generally narrow, jet like flows that travel at speeds between 40 and 120 kilometers per day. Western boundary currents are the deepest ocean surface flows, usually extending 1000 meters below the ocean surface.

Flowing from high latitudes to the equator are the eastern boundary currents. These cold water currents also have specific names associated with their location: North Atlantic - Canary; North Pacific - California; South Atlantic - Benguela; South Pacific - Peru; and Indian Ocean - West Australia. All of these currents are generally broad, shallow moving flows that travel at speeds between 3 and 7 kilometers per day.

In the Northern Hemisphere, the east flowing North Pacific Current and North Atlantic Drift move the waters of western boundary currents to the starting points of the eastern boundary currents. The South Pacific Current, South Indian Current and South Atlantic Current provide the same function in the Southern Hemisphere. These currents are associated with the Antarctic Circumpolar. Because of the absence of landmass at this latitude zone, the Antarctic Circumpolar flows in continuous fashion around Antarctica and only provides a partial return of water to the three Southern Hemispheric ocean basins.

Surface Currents of the Polar Gyres

The polar gyres exist only in the Atlantic and Pacific basins in Northern Hemisphere. They are propelled by the counterclockwise winds associated with the development of permanent low pressure centers at 50° of latitude over the ocean basins. Note that the bottom west flowing current of the polar gyres is the topmost flowing current of the subtropical gyres.

The world's oceans also have significant currents that flow beneath the surface. Subsurface currents generally travel at a much slower speed when compared to surface flows. The subsurface currents are driven by differences in the density of seawater. The density of seawater deviates in the oceans because of variations in temperature and salinity. Near surface seawater begins its travel deep into the ocean in the North Atlantic.

The downwelling of this water is caused by high levels of evaporation which cools and increases the salinity of the seawater located here. This seawater then moves south along the coast of North and South America until it reaches Antarctica. At Antarctica, the cold and dense seawater then travels eastward. During this part of its voyage the flow splits off into two currents that move northward. In the North Pacific (off the coast of Asia) and in the Indian Ocean (off the coast of Africa), these two currents move from the ocean floor to its surface creating upwellings. The flow then becomes near surface moving back to the starting point in the North Atlantic. One complete circuit of this flow of seawater is estimated to take about 1,000 years.

Ocean Tides

An ocean tide refers to the cyclic rise and fall of seawater. Tides are caused by slight variations in gravitational attraction between the Earth and the moon and the sun in geometric relationship with locations on the Earth's surface.

Tides are periodic primarily because of the cyclical influence of the Earth's rotation.

The moon is the primary factor controlling the temporal rhythm and height of tides. The moon produces two tidal bulges somewhere on the Earth through the effects of gravitational attraction. The height of these tidal bulges is controlled by the moon's gravitational force and the Earth's gravity pulling the water back toward the Earth. At the location on the Earth closest to the moon, seawater is drawn toward the moon because of the greater strength of gravitational attraction.

On the opposite side of the Earth, another tidal bulge is produced away from the moon. However, this bulge is due to the fact that at this point on the Earth the force of the moon's gravity is at its weakest. Considering this information, any given point on the Earth's surface should experience two tidal crests and two tidal troughs during each tidal period.

The timing of tidal events is related to the Earth's rotation and the revolution of the moon around the Earth. If the moon was stationary in space, the tidal cycle would be 24 hours long. However, the moon is in motion revolving around the Earth. One revolution takes about 27 days and adds about 50 minutes to the tidal cycle. As a result, the tidal period is 24 hours and 50 minutes in length.

The second factor controlling tides on the Earth's surface is the sun's gravity. The height of the average solar tide is about 50 % the average lunar tide. At certain times during the moon's revolution around the Earth, the direction of its gravitational attraction is aligned with the sun's (Figure 8). During these times the two tide producing bodies act together to create the highest and lowest tides of the year.

These spring tides occur every 14-15 days during full and new moons.

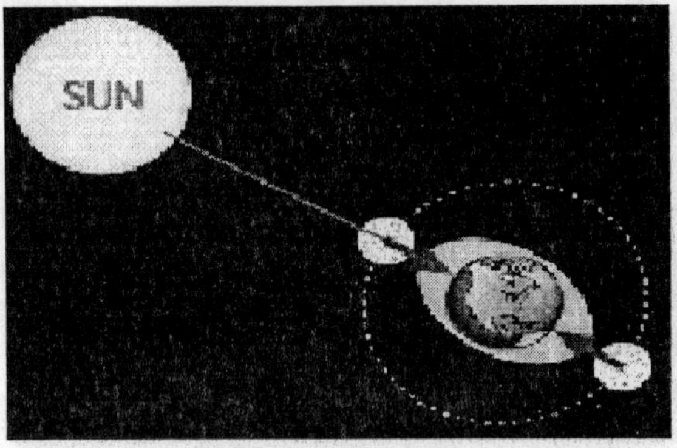

Figure 8: Forces involved in the formation of a spring tide.

When the gravitational pull of the moon and sun are at right angles to each other, the daily tidal variations on the Earth are at their least. These events are called neap tides and they occur during the first and last quarter of the moon.

Types of Tides

The geometric relationship of moon and sun to locations on the Earth's surface results in creation of three different types of tides. In parts of the northern Gulf of Mexico and Southeast Asia, tides have one high and one low water per tidal day (Figure 9). These tides are called diurnal tides.

Semi-diurnal tides have two high and two low waters per tidal day (Figure 10). They are common on the Atlantic coasts of the United States and Europe.

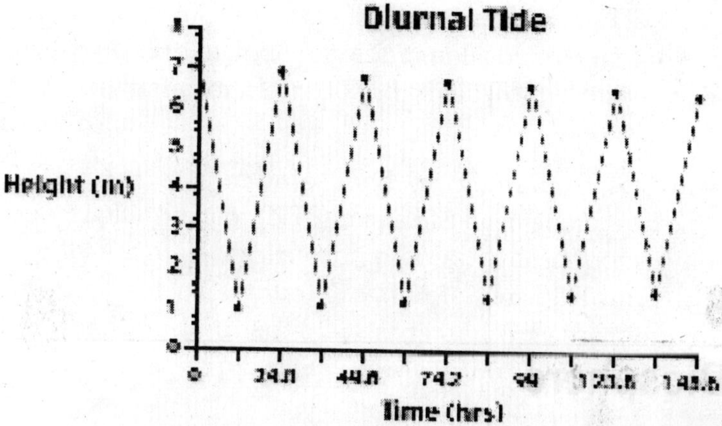

Figure 9: Cyclical tidal cycles associated with a diurnal tide

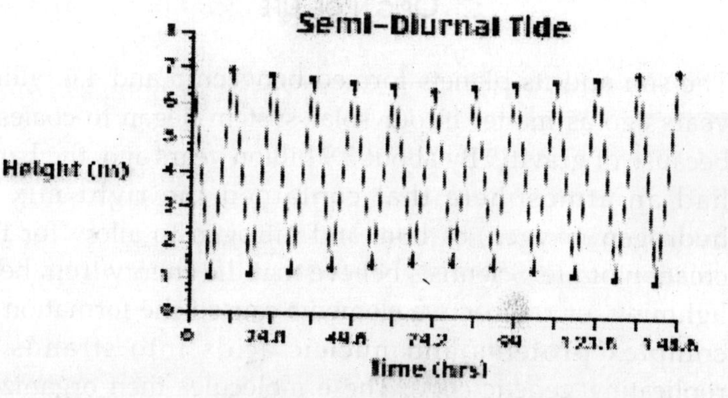

Figure 10: Cyclical tidal cycles associated with a semi-diurnal tide.

Many parts of the world experience mixed tides where successive high-water and low-water stands differ appreciably. In these tides, we have a higher high water and lower high water as well as higher low water and lower low water. The tides around west coast of Canada and the United States are of this type.

6
Biosphere

Origin of Life

The sun and its planets formed between 5 and 4.6 billion years ago as matter in our solar system began to coalesce because of gravity. By about 3.9 billion years ago, the Earth had an atmosphere that contained the right mix of hydrogen, oxygen, carbon, and nitrogen to allow for the creation of life. Scientists believe that the energy from heat, lightning, or radioactive elements caused the formation of complex proteins and nucleic acids into strands of replicating genetic code. These molecules then organized and evolved to form the first simple forms of life.

At 3.8 billion years ago, conditions became right for the fossilization of the Earth's early cellular life forms. These fossilized cells resemble present day cyanobacteria. Such cells are known as prokaryotes. Prokaryote cells are very simple, containing few specialized cellular structures and their DNA is not surrounded by a membranous envelope. The more complex cells of animals and plants, known as eukaryotes, first showed up about 2.1 billion year ago. Eukaryotes have a membrane-bound nucleus and many specialized structures located within their cell boundary.

By 680 million years ago, eukaryotic cells were beginning to organize themselves into multicellular organisms. Starting at about 570 million years ago an enormous diversification of multicellular life occurred known as the Cambrian explosion. During this period all but one modern phylum of animal life made its first appearance on the Earth. Table 1 describes the approximate time of origin of the Earth' s major groups of plants and animals.

Table 1: Time of origin of some major plant and animal groups.

Organism Group	Time of Origin
Marine Invertebrates	570 Million Years Ago
Fish	505 Million Years Ago
Land Plants	438 Million Years Ago
Amphibians	408 Million Years Ago
Reptiles	320 Million Years Ago
Mammals	208 Million Years Ago
Flowering Plants (Angiosperms)	140 Million Years Ago
Hominid Line Begins Its Evolution	20-15 Million Years Ago

Scientists currently recognize four groups of biological entities:

— *Archaea* - are a group of recently discovered organisms that sometimes live in extremely hostile habitats like thermal volcanic vents, saline pools, and hot springs. Archaea are single-celled organisms that are similar in appearance to bacteria. However, they are biochemically and genetically very different from bacteria. Many books and other forms of scientific literature refer to them as archaebacteria.

— *Bacteria* - are simple single-celled organisms that generally lack chlorophyll. Bacteria have a prokaryote cell type. They also generally obtain energy for survival through the breakdown of organic matter via fermentation and respiration. Bacteria such as Rhizobium spp. and cyanobacteria play an important role in the fixing of atmospheric nitrogen. Without these bacteria ecosystems would be severely short of nitrogen for plant and animal growth. The oldest fossils of life on Earth are bacteria-like organisms.

— *Eukaryota* - are all organisms with a eukaryote cell type. This group of life includes the kingdoms Protista, Fungi, Animalia, and Plantae.

— *Viruses* - are fragments of DNA or RNA that depend on host cells that they infect for their reproduction. They are not cells. Viruses are thought to be parts of the genetic code that originated from either eukaryote or prokaryote cells. These code fragments contain enough genetic information for self-existence. At times, viruses are metabolically inert and technically non-living. Viruses cause a variety of diseases in eukaryote organisms. In humans they can cause smallpox, chicken pox, influenza, shingles, herpes, polio, ebola, AIDS, rabies, and some types of cancer.

The four main types of biological entities described above share some unique characteristics that can allow us to distinguish them from non-living things. These characteristics are:

(1). Organisms tend to be complex and highly organized. Chemicals found within their bodies are synthesized through metabolic processes into structures that have defined purposes. Cells and their various organelles are examples of such structures. Cells are also the basic functioning unit of life. Cells are often organized into organs to create higher levels of complexity and function.

(2). Living things have the ability to take energy from their environment and change it from one form to another. This energy is usually used to facilitatc their growth and reproduction.

(3). Organisms tend to be homeostatic. In other words, they regulate their bodies and other internal structures to certain normal parameters.

(4). Living creatures respond to stimuli. Cues in their environment cause them to react through behavior, metabolism, and physiological change.

(5). Living things reproduce themselves by making copies of themselves. Reproduction can either be sexual or asexual. Sexual reproduction involves the fusing of haploid genetic material from two individuals. This process creates populations with much greater genetic diversity.

(6). Organisms tend to grow and develop. Growth involves the conversion of consumed materials into biomass, new individuals, and waste.

(7). Life adapts and evolves in step with external changes in the environment through mutation and natural selection. This process acts over relatively long periods of time.

Biological Classification of Organisms

The term species has its origins in the ancient Latin language. In this language, the word species means kind. A more technical definition of species is a group of interbreeding organisms that do not ordinarily breed with members of other groups. Biologists estimate that about 10 to 40 million different species inhabit the Earth. Of these species, approximately 1.5 million have been classified.

The first individual to propose an orderly system for classifying the variety of organisms found on our planet was Linnaeus. In his system of classification, the finest unit in the organization of life is the species. Linnaeus suggested that every organism should be classified with a unique binomial name. The first term in this classification system is the organism's generic name or Genus. The second term is the organism's specific name or species designation.

Current classification systems have developed from Linnaeus' original work. However, modern classification systems are much more complicated having many levels of hierarchical organization. These systems are also taxonomic (structural and physiological connections between organisms), phylogenic (classification based on genetic connections between organisms), and are structurally based on Darwin's theory of evolution.

Modern classifications of organisms are standardized in a hierarchical system that go from general to specific. Table 2 below describes the detailed classification of the tree red maple. Note that each level of organization is based on some biological characteristic that the organism possesses.

Table 2: Hierarchical system of the biological classification of an organism.

Category	Name	Characteristics
Kingdom	Plantae	Organisms that usually have rigid cell walls and usually possess chlorophyll.
Subkingdom	Embryophyta embryos.	Plants forming
Phylum	Tracheophyta	Vascular plants.
Subphylum	Pterophytina	Generally large, conspicuous leaves, complex vascular system.
Class	Angiospermae	Flowering plants, seed enclosed in ovary.
Subclass	Dicotyledoneae	Embryo with two seed leaves.
Order	Sapindales	Soapberry order consisting of a number of trees and shrubs.
Family	Aceraceae	Maple family.
Genus	Acer	Maples and box elder.
Species	*Acer rubrum*	Red maple.

Natural Selection and Evolution

In 1859, Charles Darwin published the book *The Origin of Species* that presented a theory to explain how life's diversity came to be. It has been suggested that without this theory nothing would make much sense in the field of modern biology.

Evolution describes the process by which species come to possess genetic adaptations to their environment. Its mechanism is natural selection. Natural selection acts through individuals by determining which individuals have the best adaptations to guarantee reproductive success. Because the state of environment is always changing temporally, natural selection is always influencing the genetic characteristics of the population over time. Thus, natural selection acts to adapt the population to its ever changing surrounds.

Evolutionary change is also a change in gene frequency. The pattern of genetic variation changes from one generation to the next as natural selection determines which individuals are fittest. New genes enter the species gene pool by way of mutations. By selecting those organisms that will become reproductively successful, natural selection controls the future frequency of a population's genes. The appearance of new mutations in a population together with the change in gene frequency results in evolution.

The spatial isolation of sub-populations from a main population is also an important condition for evolution. If a remote sub-population cannot trade genes to the main population because of isolation, natural selection acts differently in each population resulting in divergent evolution. Differences in gene frequency emerge between the two populations because no two patches of habitat are absolutely identical, and because new adaptations enter the gene pool through mutations at the level of the individual. Isolation can be due to distance, mountain barriers, a river, etc.

The best known example of natural selection operating in a modern species is the development of pesticide resistance in many insect species. Prior to the extensive use of pesticides that began in the 1940s, crop pest insect species only contained a small amount of genetic variability for resistance to these chemicals. Natural selection in the absence of pesticides could not lead to changes in the frequencies of genes causing resistance to chemical pesticides.

However, once the spraying of pesticides started, individuals that happened to possess resistant genes became much more frequent as they survived the applications. Further, they were able to pass a greater percentage of their genes on to the next generation's gene pool. As a result, the introduction of pesticides into the environment provided a tremendous selective pressure to increase the frequency of resistant genes in the pest populations.

Organization of Life: Species, Populations, Communities, and Ecosystems

Scientists have recognized that life can be organized into several different levels of function and complexity. These functional levels are: species, populations, communities, and ecosystems.

Species

Species are the different kinds of organisms found on the Earth. A more exact definition of species is a group of interbreeding organisms that do not ordinarily breed with members of other groups. If a species interbreeds freely with other species, it would no longer be a distinctive kind of organism. This definition works well with animals. However, in some plant species fertile crossings can take place among morphologically and physiologically different kinds of vegetation. In this situation, the definition of species given here is not appropriate.

Populations

A population comprises all the individuals of a given species in a specific area or region at a certain time. Its significance is more than that of a number of individuals because not all individuals are identical. Populations contain genetic variation within themselves and between other populations. Even fundamental genetic characteristics such as hair color or size may differ slightly from individual to individual. More importantly, not all members of the population are equal in their ability to survive and reproduce.

Communities

Community refers to all the populations in a specific area or region at a certain time. Its structure involves many types of interactions among species. Some of these involve the acquisition and use of food, space, or other environmental resources. Others involve nutrient cycling through all members of the community and mutual regulation of population sizes. In all of these cases, the structured interactions of populations lead to situations in which individuals are thrown into life or death struggles.

In general, ecologists believe that a community that has a high diversity is more complex and stable than a community that has a low diversity. This theory is founded on the observation that the food webs of communities of high diversity are more interconnected. Greater interconnectivity causes these systems to be more resilient to disturbance. If a species is removed, those species that relied on it for food have the option to switch to many other species that occupy a similar role in that ecosystem. In a low diversity ecosystem, possible substitutes for food may be non-existent or limited in abundance.

Ecosystems

Ecosystems are dynamic entities composed of the biological community and the abiotic environment. An ecosystem's abiotic and biotic composition and structure is determined by the state of a number of interrelated environmental factors. Changes in any of these factors (for example: nutrient availability, temperature, light intensity, grazing intensity, and species population density) will result in dynamic changes to the nature of these systems. For example, a fire in the temperate deciduous forest completely changes the structure of that system. There are no longer any large trees, most of the mosses, herbs, and shrubs that occupy the forest floor are gone, and the nutrients that were stored in the biomass are quickly released into the soil, atmosphere and hydrologic system. After a short time of recovery, the community that was once large mature trees now becomes a community of grasses, herbaceous species, and tree seedlings.

Abiotic Factors and the Distribution of Species

The geographic distributions of plant and animal species are never fixed over time. Geographic ranges of organisms shift, expand, and contract. These changes are the result of two contrasting processes: colonization and establishment and localized extinction. Colonization and establishment takes place when populations expand into new areas. A number of processes can initiate this process including disturbance and abiotic environmental change. Localized extinction results in the elimination of populations from all or part of their former range. It can be caused by biotic interactions or, once again, abiotic environmental change.

Dispersal, Colonization, and the Establishment of Species

Many of the different types of organisms that inhabit the Earth have the ability to move. This movement can be accomplished either passively or actively. Active movement requires the organism to use some appendage to initiate walking, running, flying or swimming. In passive movement, the organism uses some external force to cause transit. For example, many plants use wind for seed dispersal, while oyster larvae can passively travel great distances by sea currents.

One common reason why organisms move is to disperse to new habitats. Dispersal can be defined as the movement of individuals away from others of the same species. In most cases, organisms disperse to escape the influence of their parents and siblings. However, dispersal may also involve a large element of discovery. By finding new suitable habitats individuals can increase the range of their species. A larger range makes the species better off in terms of evolution.

Plants have developed a number of different mechanisms for dispersing their offspring. Some of the common techniques include:

— The use of specialized morphological structures to aid the transport of an individual by wind;

— The employment of particular morphological structures to transport the individual by moving water;

— The production of fruit encased seeds that other organisms consume and disperse;
— Adhesion mechanisms; and
— The physical ejection of seeds.

Once dispersed, an individual can only colonize a new site if it is devoid of other organisms and if the necessary abiotic requirements and conditions exist for its survival. Sites within ecosystems become devoid of organisms through disturbance. A disturbance can be caused by predation, climate variations, earthquakes, volcanoes, fire, animal burrowing, and even the impact of a raindrop.

Often the struggle for survival does not end with colonization of an individual on a vacant site. Once colonized an individual may not be able to establish itself over the longterm because of abiotic and biotic influences. The death of the individual may occur through competitive interaction, predation or an abiotic factor like fire.

Abiotic Factors and Tolerance Limits

Most species appear to be limited in at least part of their geographic range by abiotic factors, such as temperature, moisture availability, and soil nutrients. No species is adapted to survive under all conditions found on the Earth. All species have specific limits of tolerance to physical factors that directly effect their survival or reproductive success. The portion of the abiotic factor's range of variation which a species can survive and function in is commonly defined as the tolerance range. The level within the tolerance range at which a species or population can function most efficiently is termed the optimum.

In 1840, J. Liebig suggested that organisms are generally limited by only one single physical factor that is in shortest supply relative to demand. At one time ecologists accepted this idea so completely that they called it Liebig's Law of the Minimum and tried to determine the single limiting factor that controls the growth of numerous species. However, many studies have shown that Liebig's concept is inadequate to account for the distributional limits of a large number of species. In these cases, scientists believe that complex interactions between several physical factors are responsible for distribution patterns.

Biotic Interactions and the Distribution of Species

Interacting species have a tremendous influence on the size of each other's populations. The various mechanisms for these biotic influences are quite different from the way in which abiotic factors effect the size of populations. Biotic factors also regulate the size of populations more intensely. Finally, the influence of biotic interactions can occur at two different levels. Interspecific effects are direct interactions between species, and the intraspecific effects represent interactions of individuals within a single species.

Neutralism

Neutralism is the most common type of interspecific interaction. Neither population directly affects the other. What interactions occur are slight and indirect. The simple presence of the two species should not directly affect the population level of either. An example of neutralism would be the interaction between rainbow trout and dandelions living in a mountain valley.

Competition

When two or more organisms in the same community seek

the same resource (e.g., food, water, nesting space, ground space), which is in limiting supply to the individuals seeking it, they compete with one another. If the competition is among members of the same species, it is called intraspecific. Competition among individuals of different species it is referred to as interspecific competition. Individuals in populations experience both types of competition to a greater or lesser degree.

Competition may be the result of two different processes: exploitation or interference. Competition by exploitation occurs between individuals when the indirect effects of two or more species or individuals reduce the supply of the limiting resource or resources needed for survival. The exclusion of one organism by another can only occur when the dominant organism requires less of the limiting resource to survive. Further, the dominant species must be able to reduce the quantity of the resource to some critical level with respect to the other organism. Resource exploitation, however, does not always cause the exclusion of a species from a community. It may just cause the species involved in this interaction to experience a reduction in their potential growth.

Competition by interference occurs when an individual directly prevents the physical establishment of another individual in a portion of a habitat. Established plants can preempt the invasion and colonization of other individuals by way of dense root mats, peat and litter accumulation, and mechanical abrasion.

Amensalism

Amensalism is an interaction where one species suffers and the other interacting species experiences no effect. One particular form of amensalism is allelopathy which occurs with plants. Allelopathy involves the production and release of chemical substances by one species that inhibit the growth of another. Allelopathic substances range from acids to bases to simple organic compounds. All of these substances are known under the general term: secondary substances. Secondary substances are chemicals produced by plants that seem to have no direct use in metabolism.

A good example of a secondary substance is the antibiotic juglone which is secreted by Black Walnut (Juglans nigra) trees. This substance is known to inhibit the growth of trees, shrubs, grasses, and herbs found growing near Black Walnut trees. In the chaparral vegetation of California, certain species of shrubs, notably Salvia leucophylla (mint) and Artemisia californica (sagebrush) are known to produce allelopathic substances. Often these chemicals accumulate in the soil during the dry season reducing the germination and growth of grasses and herbs in an area up to 1 to 2 meters from the secreting plants.

Mutualism

Mutualism is the name given to associations between pairs of species that bring mutual benefit. The individuals in the populations of each mutualist species grow and/or survive and/or reproduce at a higher rate when in the presence of individuals of the other species. In most ecology or biogeography textbooks mutualisms are generally underemphasized or ignored. Yet this type of interaction is an extremely widespread phenomena. For example, most rooting plants have mutualistic associations with fungal mycorrhizae. Mycorrhizae increase the capability of plant roots to absorb nutrients like nitrogen and phosphorus. In return, the roots of the host provide support and a constant supply of carbohydrates for consumption.

Mutualistic interactions between species can be of two types: symbiotic or nonsymbiotic. In a symbiotic mutualism, individuals interact physically and their relationship is biologically essential for survival. At least one member of the pair cannot live without close contact with the other. For example, the fungal-algal symbiosis that occurs in lichens. The morphological structure of a lichen is a mass of fungal hyphae that forms around a small colony of algae cells. In this mutualism, the alga produces carbohydrates and other food by products through photosynthesis and metabolism, while the fungus absorbs the required minerals and water to allow for these processes to occur. More common in nature is the nonsymbiotic mutualism. In this interaction, the mutualists live independent lives yet cannot survive without each other

Predation, Parasitism, and Pathogens

Pathogens, parasites, and predators obtain food at the expense of their hosts and prey. These processes are basic to the entire grazing food chain above the autotroph level. Predators tend to be larger than their prey and consume them from the outside. A parasite or pathogen is smaller than its host and consumes it either from the inside or from the outside of the organism.

Tigers are ambush predators that try to approach their prey as closely as possible. They often attack their prey from behind, biting its neck or throat in the capture process. It is easy to believe that the predator-prey interaction is somehow detrimental to the prey population. This idea has led to extensive efforts to control predator populations in the name of wildlife conservation. However, functional relationships between predator-prey between species, within natural ecosystems, have coevolved over long periods time creating a dynamic balance between their interacting populations. Thus, the population sizes of predator and prey species are interregulated by delicate feedback mechanisms that control the densities of both species.

A classic example of the balance between predator and prey involves the prickly pear cactus, Opuntia spp. In the 19th century, prickly pear cactus was introduced into Australia from South America. Because no Australian predator species existed to control the population size of this cactus, it quickly expanded throughout millions of acres of grazing land. The presence of the prickly pear cactus excluded cattle and sheep from grazing vegetation and caused a substantial economic hardship to farmers. A method of control of the prickly pear cactus was initiated with the introduction of Cactoblastis cactorum, a cactus eating moth from Argentina, in 1925. By 1930, densities of the prickly pear cactus were significantly reduced.

Sometimes predator species can drive their prey into localized extinction. In complex communities, this does no particular harm to the predator if several other species exist as alternative prey.

Concept of Ecological Niche

For a species to maintain its population, its individuals must survive and reproduce. Certain combinations of environmental conditions are necessary for individuals of each species to tolerate the physical environment, obtain energy and nutrients, and avoid predators. The total requirements of a species for all resources and physical conditions determine where it can live and how abundant it can be at any one place within its range. These requirements are termed abstractly the ecological niche.

G.E. Hutchinson suggested that the niche could be modeled as an imaginary space with many dimensions, in which each dimension or axis represents the range of some environmental condition or resource that is required by the species. Thus, the niche of a plant might include the range of temperatures that it can tolerate, the intensity of light required for photosynthesis, specific humidity regimes, and minimum quantities of essential soil nutrients for uptake.

A useful extension of the niche concept is the distinction between fundamental and realized niches (Figure 1). The fundamental niche of a species includes the total range of environmental conditions that are suitable for existence without the influence of interspecific competition or predation from other species. The realized niche describes that part of the fundamental niche actually occupied by the species.

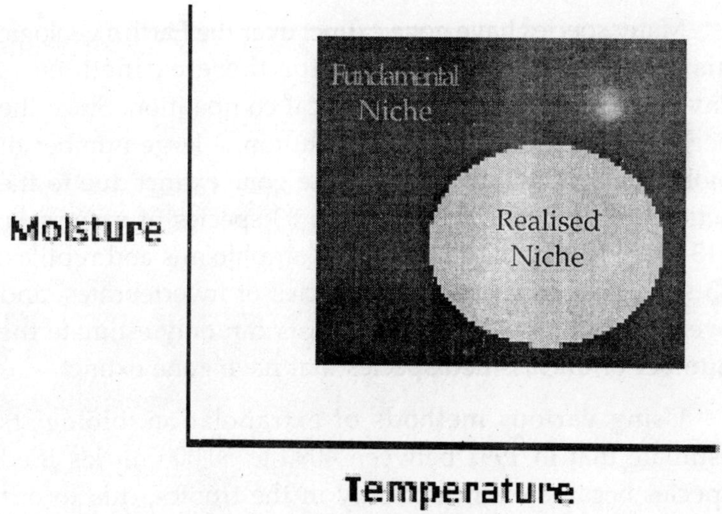

Figure 1: The diagram shows a hypothetical situation where a species distribution is controlled by just two environmental variables: temperature and moisture.

Species Diversity and Biodiversity

Biologists are not completely sure how many different species live on the Earth. Estimates of how many species exist on the Earth range from low of 2 million to high of about 100 million. To date, about 2.1 million species have been classified, primarily in the habitats of the middle latitudes. Most of the unclassified species on this planet are invertebrates. This group of organisms includes insects, spiders, mollusks, sponges, flatworms, starfish, urchins, earthworms, and crustaceans. These species are often difficult to find and identify because of their small size and the fact that they live in habitats that are difficult to explore. In the tropical rain forest, the cataloging of species has been quite limited because of this later reason. Scientists estimate that this single biome may contain 50 to 90 % of the Earth's biodiversity.

Many species have gone extinct over the Earth's geologic history. The primary reason for these extinctions is environmental change or biological competition. Since the beginning of the Industrial Revolution, a large number of biologically classified species have gone extinct due to the actions of humans. This includes 83 species of mammals, 113 species of birds, 23 species of amphibians and reptiles, 23 species of fish, about 100 species of invertebrates, and over 350 species of plants. Scientists can only estimate the number of unclassified species that have gone extinct

Using various methods of extrapolation, biologists estimate that in 1991 between 4000 to 50,000 unclassified species became extinct, mainly in the tropics, due to our activities. This rate of extinction is some 1,000 to 10,000 times greater than the natural rate of species extinction prior to the appearance of human beings. The continued extinction of species on this planet by human activities is one of the greatest environmental problems facing humankind.

Several times during the Earth's history there have been periods of mass extinctions, when many species became extinct in a relatively short time period (a few million years is a relatively short time when compared to the age of the Earth). Scientists are unsure of the causes of both background extinction and mass extinction. Possible explanations for mass extinctions include climate changes or catastrophes such as the Earth being hit by a meteor. Since the beginning of time, five or six mass extinctions have occurred that eliminated between 35 % and 96 % of all species on Earth. Further, it is believed that of all species that ever inhabited the Earth over 99 % of them are now extinct.

Assessment of the number of different organisms that live on this planet is plagued with difficulties. First and foremost, biologists lack a precise definition of what exactly defines a species. The concept of a species often refers to a population of physically similar individuals that can successfully mate between each other, but cannot produce fertile offspring with other organisms. However, many species are composed of a number of distinct populations that can interbreed even though they display physiological and anatomical differences. Scientists developed the notion of biodiversity to overcome some of the difficulties of species concept. To accomplish this task, biodiversity describes the diversity of life at the following three biological levels:

— *Genetic Level or Genetic Diversity* - Genetic diversity refers to the total number of genetic characteristics expressed and recessed in all of the individuals that comprise a particular species

— *Species Level or Species Diversity* - Species diversity is the number of different species of living things living in an area. As mentioned above, a species is a group of plants or animals that are similar and able to breed and produce viable offspring under natural conditions.

— *Ecosystem Level or Ecosystem Diversity* - Ecosystem diversity is the variation of habitats, community types, and abiotic environments present in a given area. An ecosystem consists of all living and non-living things in a given area that interact with one another.

The biodiversity found on Earth today is the product of 3.5 billion years of evolution. In fact, the Earth supports more biodiversity today than in any other period in history. However, much of this biodiversity is now facing the threat of extinction because of the actions of humans.

Plant Sussession

Succession is a directional non-seasonal cumulative change in the types of plant species that occupy a given area through time. It involves the processes of colonization, establishment, and extinction which act on the participating plant species. Most successions contain a number of stages that can be recognized by the collection of species that dominate at that point in the succession. Succession begin when an area is made partially or completely devoid of vegetation because of a disturbance. Some common mechanisms of disturbance are fires, wind storms, volcanic eruptions, logging, climate change, severe flooding, disease, and pest infestation. Succession stops when species composition changes no longer occur with time, and this community is said to be a climax community.

The concept of a climax community assumes that the plants colonizing and establishing themselves in a given region can achieve stable equilibrium. The idea that succession ends in the development of a climax community has had a long history in the fields of biogeography and ecology. One of the earliest proponents of this idea was Frederic Clements who studied succession at the beginning of the 20th century. However, beginning in the 1920s scientists began refuting the notion of a climax state. By 1950, many scientists began viewing succession as a phenomenon that rarely attains equilibrium. The reason why equilibrium is not reached is related to the nature of disturbance.

Disturbance acts on communities at a variety of spatial and temporal scales. Further, the effect of disturbance is not always 100 percent. Many disturbances remove only a part of the previous plant community. As a result of these new ideas, plant communities are now generally seen as being composed of numerous patches of various size at different stages of successional development.

One of the earliest studies of plant succession was done by Dwight Billings in the 1930s. In this investigation, Billings examined the succession of plant species that occurred on abandoned agricultural fields in North Carolina. Billings studied a number of fields that had been deserted from just a few years to a maximum of about 150 years. From observations of the plant communities that existed in these sites, Billings was able to construct a detailed successional sequence.

The first stage of succession was characterized by the pioneering colonization of annual plant species on bare ground and nutrient poor soils. These annual species had short lifespans (one growing season), rapid maturity, and produce numerous small easily dispersed seeds. The annuals were then quickly replaced in dominance in the next year by biennial plants and grasses. After about 3 to 4 years, the biennial and grass species gave way to perennial herbs and shrubs. These plants live for many years and have the ability to reproduce several times over their lifespans.

Pioneer species consist of a variety of annual plants. This successional stage is then followed by communities of perennials and grasses, shrubs, softwood trees and shrubs, and finally hardwood trees and shrubs. This succession takes about 120 years to go from the pioneer stage to the climax community. After about 5 to 15 years, the sites were then colonized by a number of different softwood tree species including loblolly pine (Pinus taeda), shortleaf pine (Pinus echinata), Virginia pine (Pinus virginiana), and sweetgum.

As the softwoods increased in their numbers and grew in height, they began forming a forest canopy. This canopy reduces the amount of light reaching the forest floor. The resulting shaded understory conditions caused the exclusion of many light loving perennial herb and shrub species. Low light conditions also inhibited the germination of pine seedlings. Perennial herb and shrub species that were adapted to low light conditions now began to take over the ground cover.

The canopy also changed the microclimate of habitat near ground level. It was now more humid, has moderated temperatures, and less wind. These conditions, plus the development of a soil litter layer, allowed for the germination of hardwood species, like oak (Quercus spp.) and various species of hickory (Carya spp). The seedlings of these tree species also tolerate low light levels.

By about 50 to 75 years after the initial colonization of the pioneer species, the hardwoods started to replace the softwood species in the developing forest. At this stage in the succession, the pines had maximum heights of about 25 meters, while the oaks and hickories were on average about 10 meters tall. Because of their shorter lifespans (50 years), many of the softwood species were beginning to die out and the gap that was created was then filled by a subdominant hardwood tree. Hardwood species, like oak and hickory, can live for more than 100 years. Sites more than 100 years old were found to be dominated by mature oak forests.

Organismic and Individualistic Views of Succession

In the first quarter of this century there was considerable debate about the nature of the community. F.E. Clements conceived of the community as a sort of superorganism whose member species were tightly bound together both now and in their common evolutionary history. Thus, individuals, populations, and communities have a relationship to each other that resembles the associations found between cells, tissues, and organs.

Clements' theories on communities were also linked to succession. His successional concept suggested that communities of organisms are subject to special laws in which the action of the whole is greater than the sum of the parts, and that this action results in a directional change in the species composition of the community to a climax state controlled primarily by climate.

The main processes acting to produce the various successional stages of species dominance, and finally climax, are competition and plant modification of the abiotic environment. Environmental modification, however, is detrimental to the plants doing the modifying. Modification changes the environment allowing the establishment of new colonists, and then results in the subsequent competitive exclusion of the former inhabitants by these colonists. This facilitative process stops when the climax community is reached.

Clements presented a deterministic unidirectional view of succession where the present pattern is governed by the past pattern. The philosophical structure of Clements' holistic approach is quite similar to advances in other sciences of that time. In geology and geomorphology theorists presented views that contained various stages of maturity, and compared landscape evolution metaphorically to a developing organism. The parallel in theoretical approach of these sciences with the work of Clements may be the result of attitudes prevalent in science as a whole at that time.

In 1926, H. Gleason devised a new theory to explain the nature of communities. Gleason's individualistic concept saw the relationship between coexisting species as simply the result of similarities in their requirements and tolerances (and partly the result of chance). Taking this view, community boundaries need not be sharp, and associations of species would be much less predictable than one would expect from the superorganism concept.

Gleason argued that the holistic view point of Clements was inadequate in explaining the mechanism of succession. For example, Gleason suggested that Clementsian concepts could not properly explain the occurrence of such phenomena as retrogressive successions. In reference to his view of succession, Gleason stated that ".... every species of plant is a law unto itself, the distribution of which in space depends upon its individual peculiarities of migration and environmental requirements". Thus, associations of plants, or communities, were not highly organized, but aggregations of independent plant species, each specialized to survive on habitats they were adapted for.

Retrogressive successions were possible in Gleason's model if environmental variables deteriorated with time, changing the pattern of establishment, growth and reproduction of plants in a habitat. Clements' model, however, assumed long term climatic stability, and this assumption does not allow for short term retrogressive community change.

Clements and Gleason presented two diametrically opposed opinions on community organization and structure. Further investigation in this discussion will show that these views are still present in the hypotheses of later theorists, but in a somewhat modified form. Many of these modified hypotheses involve a synthesis of the early ideas of Clements and Gleason. This synthesis is the result of the addition of new ecological information or the re-analysis of old information on how ecosystems function over time.

The synthetic evolution of successional hypotheses must be expected, as investigation finds new mechanisms responsible for temporal vegetation change in a relatively unexplored world. The early presence of simple diametrically opposed successional hypotheses in the early years is probably the result of the immature state of understanding of turn of the century ecology.

Central to Gleason's succession model is the notion of abiotic and biotic heterogeneity in space and time. This concept is a characteristic view of much of modern ecology. Recently, several scientists have examined the role of disturbance on community structure. These researchers suggest that disturbance is a common process in most communities that shapes the nature and structure of biotic interactions and processes. These ideas follow directly from Gleason's early observations of pattern and process in the plant community.

The individualistic concept of succession outlined by Gleason was ignored by the scientific community for some twenty to thirty years. Results of many studies indicate that a given location, by virtue mainly of its physical characteristics, possesses a reasonable predictable association of species. However, a given species that occurs in one predictable association is also quite likely to occur with another group of species under different conditions elsewhere.

Types of Succession

— *Primary succession* - is the establishment of plants on land that has not been previously vegetated - Mount Saint Helens. Begins with colonization and establishment of pioneer species.

— *Secondary succession* - is the invasion of a habitat by plants on land that was previously vegetated. Removal of past vegetation may be caused by natural or human disturbances such as fire, logging, cultivation, or hurricane.

— *Allogenic succession* - is caused by a change in environmental conditions which in turn influences the composition of the plant community. There have been a number of reports of an apparent allogenic transition between salt-marsh and woodland. The estuary of the River Fal in Cornwall, England, like many other estuaries, is subject to a quite rapid deposition of silt. This accretion can occur at a rate of 1 cm per year on the mud flats which are found 15 kilometers into the estuary. As a result, salt marsh has extended 800 meters seawards during the last century, while valley woodland has kept pace by invading the landward limit of the marshland.

— *Autogenic succession* - is a succession where both the plant community and environment change, and this change is caused by the activities of the plants over time. Mt. St. Helens after the last volcanic eruption.

— *Progressive succession* - is a succession where the community becomes complex and contains more species and biomass over time.

— *Retrogressive succession* - is a succession where the community becomes simplistic and contains fewer species and less biomass over time. Some retrogressive successions are allogenic in nature. For example, the introduction of grazing animals result in degenerated rangeland.

Table 3 describes some of the plant, community, and ecosystem attributes that change with succession.

Table 3: Comparison of plant, community, and ecosystem characteristics between early and late stages of succession

Attribute	Early Stages of Succession	Late Stages of Succession
Plant Biomass	Small	Large
Plant Longevity	Short	Long
Seed Dispersal Characteristics of Dominant Plants	Well dispersed	Poorly dispersed
Plant Morphology and Physiology	Simple	Complex
Photosynthetic Efficiency of		

Dominant Plants at Low Light	Low	High
Rate of Soil Nutrient Resource Consumption by Plants	Fast	Slow
Plant Recovery Rate from Resource Limitation	Fast	Slow
Plant Leaf Canopy Structure	Multilayered	Monolayer
Site of Nutrient Storage and Litter	Litter and Soil	Living Biomass
Role of Decomposers in Cycling Nutrients to Plants	Minor	Great
Biogeochemical Cycling	Open and Rapid	Closed and Slow
Rate of Net Primary Productivity	High	Low
Community Site Characteristics (Mesic)	Extreme	Moderate
Importance of Macroenvironment on Plant Success	Great	Moderate
Ecosystem Stability	Low	High
Plant Species Diversity	Low	High
Life-History Type	r	K
Seed Longevity	Long	Short

Succession Mechanisms

An overview of the mechanisms of succession has been produced by Connell and Slatyer. Connell and Slatyer propose three models, of which the first (facilitation) is the classical explanation most often invoked in the past, while the other two (tolerance and inhibition) may be equally important but have frequently been overlooked.

The essential feature of facilitation succession, in contrast with either the tolerance or inhibition models, is that changes in the abiotic environment are imposed by the developing plant community. Thus, the entry and growth of the later species depends on earlier species preparing the ground.

The tolerance model suggests that a predictable sequence is produced because different species have different strategies for exploiting resources. Later species are able to tolerate lower resource levels due to competition and can grow to maturity in the presence of early species, eventually out competing them.

The inhibition model applies when all species resist invasions of competitors. Later species gradually accumulate by replacing early individuals when they die. An important distinction between models is the cause of death of the early colonists. In the case of facilitation and tolerance, they are killed in competition for resources, notably light and nutrients. In the case of the inhibition model, however, the early species are killed by very local disturbances caused by extreme physical conditions or the action of predators.

Ecosystem Concept

An ecosystem was defined as a dynamic entity composed of a biological community and its associated abiotic environment. Often the dynamic interactions that occur within an ecosystem are numerous and complex. Ecosystems are also always undergoing alterations to their biotic and abiotic components. Some of these alterations begin first with a change in the state of one component of the ecosystem which then cascades and sometimes amplifies into other components because of relationships.

In recent years, the impact of humans has caused a number of dramatic changes to a variety of ecosystems found on the Earth. Humans use and modify natural ecosystems through agriculture, forestry, recreation, urbanization, and industry. The most obvious impact of humans on ecosystems is the loss of biodiversity. The number of extinctions caused by human domination of ecosystems has been steadily increasing since the start of the Industrial Revolution.

The frequency of species extinctions is correlated to the size of human population on the Earth which is directly related to resource consumption, land-use change, and environmental degradation. Other human impacts to ecosystems include species invasions to new habitats, changes to the abundance and dominance of species in communities, modification of biogeochemical cycles, modification of hydrologic cycling, pollution, and climatic change.

Components of Ecosystems

Ecosystems are composed of a variety of abiotic and biotic components that function in an interrelated fashion. Some of the more important components are: soil, atmosphere, radiation from the sun, water, and living organisms.

Soils are much more complex than simple sediments. They contain a mixture of weathered rock fragments, highly altered soil mineral particles, organic matter, and living organisms. Soils provide nutrients, water, a home, and a structural growing medium for organisms. The vegetation found growing on top of a soil is closely linked to this component of an ecosystem through nutrient cycling.

The atmosphere provides organisms found within ecosystems with carbon dioxide for photosynthesis and oxygen for respiration. The processes of evaporation, transpiration, and precipitation cycle water between the atmosphere and the Earth's surface.

Solar radiation is used in ecosystems to heat the atmosphere and to evaporate and transpire water into the atmosphere. Sunlight is also necessary for photosynthesis. Photosynthesis provides the energy for plant growth and metabolism, and the organic food for other forms of life.

Most living tissue is composed of a very high percentage of water, up to and even exceeding 90 %. The protoplasm of a very few cells can survive if their water content drops below 10 %, and most are killed if it is less than 30-50 %. Water is the medium by which mineral nutrients enter and are translocated in plants. It is also necessary for the maintenance of leaf turgidity and is required for photosynthetic chemical reactions. Plants and animals receive their water from the Earth's surface and soil. The original source of this water is precipitation from the atmosphere.

Ecosystems are composed of a variety of living organisms that can be classified as producers, consumers, or decomposers. Producers or autotrophs, are organisms that can manufacture the organic compounds they use as sources of energy and nutrients. Most producers are green plants that can manufacture their food through the process of photosynthesis. Consumers or heterotrophs get their energy and nutrients by feeding directly or indirectly on producers. We can distinguish two main types of consumers. Herbivores are consumers that eat plants for their energy and nutrients.

Organisms that feed on herbivores are called carnivores. Carnivores can also consume other carnivores. Plants and animals supply organic matter to the soil system through shed tissues and death. Consumer organisms that feed on this organic matter, or detritus, are known as detritivores or decomposers. The organic matter that is consumed by the detritivores is eventually converted back into inorganic nutrients in the soil. These nutrients can then be used by plants for the production of organic compounds.

Energy and Matter Flow in Ecosystems

Many of the most important relationships between living organisms and the environment are controlled ultimately by the amount of available incoming energy received at the Earth's surface from the sun. It·is this energy which helps to drive biotic systems. The sun's energy allows plants to convert inorganic chemicals into organic compounds.

Only a very small proportion of the sunlight received at the Earth's surface is transformed into biochemical form. Several studies have been carried out to determine this amount. A study of an Illinois cornfield reported that 1.6 % of the available solar radiation was photosythetically utilized by the corn. Other data suggests that even the most efficient ecosystems seldom incorporate more than 3 % of the available solar insolation. Most ecosystems fix less than 1 % of the sunlight available for photosynthesis.

Living organisms can use energy in basically two forms: radiant or fixed. Radiant energy exists in the form of electromagnetic energy, such as light. Fixed energy is the potential chemical energy found in organic substances. This energy can be released through respiration. Organisms that can take energy from inorganic sources and fix it into energy rich organic molecules are called autotrophs. If this energy comes from light then these organisms are called photosynthetic autotrophs. In most ecosystems plants are the dominant photosynthetic autotroph.

Organisms that require fixed energy found in organic molecules for their survival are called heterotrophs. Heterotrophs who obtain their energy from living organisms are called consumers. Consumers can be of two basic types: Consumer and decomposers. Consumers that consume plants are know as herbivores. Carnivores are consumers who eat herbivores or other carnivores. Decomposers or detritivores are heterotrophs that obtain their energy either from dead organisms or from organic compounds dispersed in the environment.

Once fixed by plants, organic energy can move within the ecosystem through the consumption of living or dead organic matter. Upon decomposition the chemicals that were once organized into organic compounds are returned to their inorganic form and can be taken up by plants oncc again. Organic energy can also move from one ecosystem to another by a variety of processes. These processes include: animal migration, animal harvesting, plant harvesting, plant dispersal of seeds, leaching, and erosion.

Earth's Terrestrial Biomes

Many places on Earth share similar climatic conditions despite being found in geographically different areas. As a result of natural selection, comparable ecosystems have developed in these separated areas. Scientists call these major ecosystem types biomes. The geographical distribution (and productivity) of the various biomes is controlled primarily by the climatic variables precipitation and temperature.

Most of the classified biomes are identified by the dominant plants found in their communities. For example, grasslands are dominated by a variety of annual and perennial species of grass, while deserts are occupied by plant species that require very little water for survival or by plants that have specific adaptations to conserve or acquire water.

The diversity of animal life and subdominant plant forms characteristic of each biome is generally controlled by abiotic environmental conditions and the productivity of the dominant vegetation. In general, species diversity becomes higher with increases in net primary productivity, moisture availability, and temperature.

Adaptation and niche specialization are nicely demonstrated in the biome concept. Organisms that fill similar niches in geographically separated but similar ecosystems usually are different species that have undergone similar adaptation independently, in response to similar environmental pressures. The vegetation of California, Chile, South Africa, South Australia, Southern Italy and Greece display similar morphological and physiological characteristics because of convergent evolution. In these areas, the vegetation consists of drought-resistant, hard-leaved, low growing woody shrubs and trees like eucalyptus, olive, juniper, and mimosa.

ARCTIC AND ALPINE TUNDRA

Tundra means marshy plain. The geographical distribution of the tundra biome is largely poleward of 60° North latitude. The tundra biome is characterized by an absence of trees, the presence of dwarf plants, and a ground surface that is wet, spongy, and hummocky. Soils of this biome are usually permanently frozen (permafrost) starting at a depth of a few centimeters to meter or more. The permafrost line is a physical barrier to plant root growth.

Within this biome, temperature, precipitation, and evaporation all tend to be at a minimum. Most tundra locations, have summer months with an average temperature below 10° Celsius. Precipitation in the wettest month is normally no higher that 25 millimeters. However, despite the low levels of precipitation the ground surface of the tundra biome is often waterlogged because of low rates of evapotranspiration.

The species diversity of tundra vegetation is relatively small. Plant communities are usually composed of a few species of dwarf shrubs, a few grass species, sedges, and mosses. The principal herbivores in this biome include caribou, musk ox, arctic hare, voles, and lemmings. Most of the bird species of the tundra have the ability to migrate and live in warmer locations during the cold winter months. The herbivore species support a small number of carnivore species like the arctic fox, snow owl, polar bear, and wolves. Reptiles and amphibians are few or completely absent because of the extremely cold temperatures.

Alpine tundra is quite similar to some arctic tundra but differs in the absence of permafrost and in the presence of better drainage.

Boreal Coniferous Forest

This moist-cool, transcontinental coniferous forest, or taiga lies largely between the 45th and 57th North latitudes. The climate of this biome is cool to cold with more precipitation than the tundra, occurring mainly in the summer because of mid-latitude cyclones. The predominant vegetation of boreal biome are needle-leaf evergreen variety tree species. Some common species include: White Spruce (Picea glauca) and Balsam Fir (Abies balsamea) east of the Rockies; Red Pine (Pinus resinosa), White Pine (Pinus strobus), and Hemlock (Tsuga canadensis) in the Great Lakes Region.

In British Columbia, dominant boreal trees include: Black Spruce (Picea mariana), White Spruce (Picea glauca), Lodgepole Pine (Pinus contorta), Ponderosa Pine (Pinus pondersoa), Douglas Fir (Pseudotsuga menziesii), and Alpine Fir (Abies lasiocarpa). The understory is relatively limited as a result of the low light penetration even during the spring and fall months. Understory plants in the deciduous biome take advantage of the leafless condition of trees during these seasons concentrating there growth during this time period. Common understory species include orchids, shrubs like rose, blueberry, and cranberry.

Mammals common to the boreal forest include moose, bear, deer, wolverine, marten, lynx, wolf, snowshoe hare, vole, chipmunks, shrews, and bats. Reptiles are rare, once again, because of cold temperatures. Boreal forest soils are characterized by a deep litter layer and slow decomposition. Soils of this biome are also acidic and mineral deficient because of the large movement of water vertically though the profile and subsequent leaching.

Temperate Deciduous Forest

As its name indicates, this biome is characterized by a

moderate climate and deciduous trees. It once occupied much of the eastern half of the United States, central Europe, Korea, and China. This biome has been very extensively affected by human activity, and much of it has been converted into agricultural fields or urban developments. Dominant plants include trees like Maple (Acer spp.), Beech (Fagus spp.), Oak (Quercus spp.), Hickory (Carya spp.), Basswood (Tilia spp.), Cottonwood (Populus spp.), Elm (Ulmus spp.), and Willow (Salix spp.). The understory of shrubs and herbs in a mature deciduous forest is typically well developed and richly diversified.

Brown forest soils characterize temperate deciduous forest ecosystems. The surface litter layer in these soils is thin due to rapid decomposition.

Grassland

In central North America are the grasslands, the tall grass prairie toward the east and the short grass prairie westward. In Europe and Asia some grasslands are called Steppes. In South America, grasslands are known as Pampas. Prior to modern man, the tall grass prairie was dominated by species of Bluestem. This particular species dominated much of the tall grass prairie forming dense covers 1.5 to 2.0 meters tall. In the western end of the prairie, where precipitation is less, Buffalo Grass (Buchloe dactyloides) and other grasses only a few inches above the soil surface are common in this habitat. Flowering herbs, including many kinds of composites and legumes, are common but much less important than grass species. Trees are limited to low lying areas and the narrow zone immediately adjacent to streams.

In the tall grass prairie organic rich and black chernozemic soils are common. Chernozems are among the richest in nutrients and consequently the most fertile in the world. In drier parts of prairies, soils can be influenced by salinization. As a result of their fertility, most grassland ecosystems have been modified by humans to grow grain and other dryland crops.

Grassland mammals are dominated by smaller burrowing herbivores (prairie dogs, jack rabbits, ground squirrels, and gophers) and larger running herbivores such as bison, pronghorn antelope, and elk. Carnivores include badger, coyote, ferret, wolf, and cougar. The population size of many of these species has been drastically reduced because of habitat destruction. Some of these species are on the edge of extinction.

Desert

In its most typical form, the desert consists of shrub covered land where the plants are spatially quite dispersed. In general, the major desert biomes of the Earth are geographically found at between 25 to 40° North and South latitude, in the interiors of continents. Climatically, deserts are influenced by descending air currents which limit the formation of precipitation.

Many desert areas have less than 250 millimeters of precipitation annually. Dominant plants include drought resistant shrubs like the Creosote Bush (Larrea divaricata) and Sagebrush (Artemisia tridentata), water storing succulents like cactus, and many species are short lived annuals that complete their life cycles during infrequent and short rainy periods. Desert habitats can be devoid of vegetation if precipitation is in very short supply.

Most desert mammals tend to be nocturnal to avoid the high temperatures. Desert habitats have a rich lizard and snake fauna because high temperatures promote the success of cold blooded life forms.

Because productivity is low, the litter layer is comparably limited and organic content of surface soil layers is very low. Also, evaporation tends to concentrate salts at the soil surface.

Chaparral

Chaparral has a very specific spatial distribution. It is found in a narrow zone between 32 and 40° latitude North and South on the west coasts of the continents. This area has a dry climate because of the dominance of the subtropical high pressure zone during the fall, summer, and spring months.

Precipitation falls mainly in the winter months because of the seasonal movement of the polar front and its associated mid-latitude cyclone storms. Annual averages range from about 300 to 750 millimeters and most of this rain falls in a period between 2 to 4 months long. As a result of the climate, the vegetation that inhabits this biome exhibits a number of adaptations to withstand drought and fire. Trees and shrubs living in this zone tend to be small with hard evergreen leaves. Plants in the chaparral do not drop their leaves during the dry season because of the expense of replacement.

The dry climate slows the rate of leaf decomposition in the soil. As a result, the plants growing in this biome do not have nutrients available for uptake to produce new leaves when the wet season begins. Instead, the plants of the chaparral develop leaves that can withstand arid conditions.

Representative species of the chaparral include cork oak (Quercus suber), olive (Olea europaea), eucalyptus, arbutus (Arbutus unedo), acacia, maritime pine (Pinus pinaster), shrub oak (Quercus dumosa), and live oak (Quercus agrifolia). Many of the plant species have thorns to protect them from herbivore damage.

This biome is sometimes also called Mediterranean Scrubland or sclerophyll forest.

Tropical Savanna

Tropical savannas are grasslands with scattered drought-resistant trees that generally do not exceed 10 meters in height. Tree and shrub species in the savanna usually shed their leaves during the dry season. This adaptation reduces water loss from the plants. New leaves appear several weeks before the start of the rain season. Scientists believe that savanna plant species may have developed this strategy to take advantage of the season variance of the start of the rains. Climatically, these biomes are characterized by distinct wet and dry seasons. Temperatures are hot all year long.

The savanna biome constitutes extensive areas in eastern Africa, South America, and Australia. Savannas also support the richest diversity of grazing mammals in the world. The grazing animals provide food for a great variety of predators. The soils are more nutrient rich than tropical forest soils. Some soils become extremely dry because of evaporation and form laterite layers.

Tropical Rainforest

Tropical rainforests occur in a broad zone outside the equator. Annual rainfall, which exceeds 2000 to 2250 millimeters, is generally evenly distributed throughout the year. Temperature and humidity are relatively high through the year. Flora is highly diverse: a square kilometer may contain as many as 100 different tree species as compared to 3 or 4 in the temperate zone. The various trees of the tropical rain forests are closely spaced together and form a thick continuous canopy some 25 to 35 meters tall. Every so often this canopy is interrupted by the presence of very tall trees (up to 40 meters) that have wide buttressed bases for support. Epiphytic orchids and bromeliads, as well as vines (lianas), are very characteristic of the tropical rainforest biome. Some other common plants include ferns and palms. Most plants are evergreen with large, dark green, leathery leaves.

The tropical rainforest is also home to a great variety of animals. Some scientists believe that 30 to 50 % of all of the Earth's animal species may be found in this biome. Decomposition is rapid in the tropicals because high temperatures and an abundance of moisture. Because of the frequent and heavy rains, tropical soils are subject to extreme chemical weathering and leaching. These environmental conditions also make tropical soils acidic and nutrient poor.

PRIMARY PRODUCTIVITY OF PLANTS

The bodies of living organisms within a unit area make up a standing crop of biomass. More specifically, biomass can be defined as the mass of organisms per unit area and is usually expressed in units of energy or dry organic matter. Most of the biomass in a community is composed of plants, which are the primary producers of biomass because of their ability to fix carbon through photosynthesis. This chemical reaction can be described by the following simple formula:

$$6CO_2 + 6H_2O + \text{light energy} >>> C_6H_{12}O_6 + 6O_2$$

The product of photosynthesis is a carbohydrate, such as the sugar glucose, and oxygen which is released into the atmosphere. All of the sugar produced in the photosynthetic cells of plants and other organisms is derived from the initial chemical combining of carbon dioxide and water with sunlight.

This chemical reaction is catalyzed by chlorophyll acting together with other pigment, lipid, sugar, protein, and nucleic acid molecules. Sugars created in photosynthesis can be later converted by the plant to starch for storage, or it can be combined with other sugar molecules to form specialized carbohydrates, such as cellulose. Sugars can also be combined with other nutrients such as nitrogen, phosphorus, and sulfur, to build complex molecules such as proteins and nucleic acids.

Inputs and outputs of the photosynthetic process

The primary productivity of a community is the amount of biomass produced through photosynthesis per unit area and time by plants, the primary producers. Primary productivity is usually expressed in units of energy or in units of dry organic matter. Globally, primary production amounts to 243 billion metric tons of dry plant biomass per year.

The total energy fixed by plants in a community through photosynthesis is referred to as gross primary productivity (GPP). Because all the energy fixed by the plant is converted into sugar, it is theoretically possible to determine a plant's energy uptake by measuring the amount of sugar produced. A proportion of the energy of gross primary productivity is used by plants in a process called respiration. Respiration provides a plant with the energy needed for various plant physiological and morphological activities. The general equation for respiration is:

$$C_6H_{12}O_6 + 6O_2 >>> 6CO_2 + 6H_2O + \text{released energy}$$

Subtracting respiration from gross primary production gives us net primary productivity (NPP), which represents the rate of production of biomass that is available for consumption (herbivory) by heterotrophic organisms (bacteria, fungi, and animals).

Globally, patterns of primary productivity vary both spatially and temporally. The least productive ecosystems are those limited by heat energy and water like the deserts and the polar tundra. The most productive ecosystems are systems with high temperatures, plenty of water and lots of available soil nitrogen.

Table 4 describes the approximate average net primary productivity for a variety of ecosystem types.

Table 4: Average annual Net Primary Productivity of the Earth's major biomes.

Ecosystem Type	Net Primary Productivity (kilocalories/meter -2 /year)
Tropical Rain Forest	9000
Estuary	9000
Swamps and Marshes	9000
Savanna	3000
Deciduous Temperate Forest	6000
Boreal Forest	3500
Temperate Grassland	2000
Polar Tundra	600
Desert	< 200

Factors Limiting Primary Productivity

Although all biological activity in plants is ultimately dependent on received solar radiation, it is obvious that solar radiation alone does not determine primary productivity. All plants require sunlight, carbon dioxide, and water for photosynthesis.

Photosynthesis is also dependent on temperature and soil nutrients. Temperature (heat) controls the rate of plant metabolism which in turn determines the amount of photosynthesis that can take place. Most biological metabolic activity takes place within the range 0 to 50° Celsius. There is little activity above or below this range. The optimal temperatures for productivity coincide with 15 to 25° Celsius optimal range of photosynthesis.

Water is a principal requirement for photosynthesis and the main chemical component of most plant cells. In dry regions, there is a linear increase in NPP with increased water availability. In the more humid forest climates of the world, plant productivity begins to level off at higher levels of precipitation.

Efficiency of Solar Radiation Use

Depending on location, between 0 and 5 joules of solar energy is received on each square meter of the Earth's surface every minute. However, only a small proportion of this energy is converted by photosynthesis into plant biomass. Of the sunlight being received at the Earth's surface, only about 44 % of the incident shortwave radiation occurs at wavelengths (known as photosynthetically active radiation, PAR) useful for photosynthesis.

Yet, even the most efficient plant species (most of which are crop plants) can only incorporate 3-10 % of photosynthetically active radiation into the production of biomass. Of the Earth's various biomes, tropical rainforest and conifer forest are the most efficient converting between

1 to 3 % of the useable solar radiation into biomass. Dcciduous forests achieve photosynthetic efficiencies between 0.5 to 1.0 %. The desert biome has the lowest efficiencey of solar radiation use. The plants in this biome convert only 0.01 to 0.2 % of photosynthetically active radiation to biomass.

Production by Consumers and the Grazing Food Chain

Biological communities also include organisms that consume biomass for their nutrition. Such organisms include herbivores, carnivores, and detritivores. These organisms obtain their energy through respiration, a process that releases energy from organic molecules like glucose. The equation for respiration is described below:

$$C_6H_{12}O_6 + 6O_2 = 6CO_2 + 6H_2O + \text{released energy}$$

The amount of energy actually used by these animal populations is significantly less than the amount consumed. In all animals, digestion is an imperfect process and only a portion of the energy ingested is actually assimilated and then used for body maintenance, growth and reproduction. The remaining portion leaves the organism's body undigested.

We sometimes call the energy assimilated by consumer organisms the gross secondary productivity. Gross secondary productivity can be determined directly, unlike gross primary productivity, by measuring the amount consumed minus the material defecated.

The Grazing Food Chain

The grazing food chain is a model that describes the general flow of energy in communities. For most ecosystems the model begins with the photosynthetic fixation of light, carbon dioxide, and water by plant autotrophs (primary producers) who produce sugars and other organic molecules. Once produced, these compounds can be used to create the various types of plant tissues.

Primary consumers or herbivores form the second link in the grazing food chain. They gain their energy by consuming primary producers. Secondary consumers or primary carnivores, the third link in the chain, gain their energy by consuming herbivores. Tertiary consumers or secondary carnivores are animals that receive their organic energy by consuming primary carnivores. The illustation below models this process:

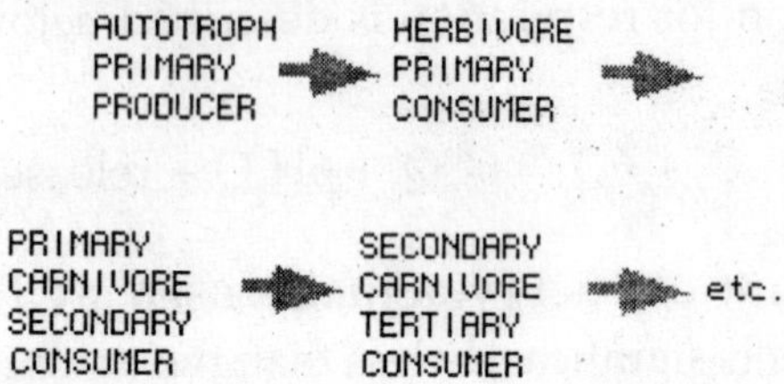

The various levels in the grazing food chain are linked to each other like links in a chain. The levels are often called trophic levels, and they suggest a particular order for the passage of energy through the food chain. Like many very simple models, the idea of a food chain only provides a simple abstraction of the nature of energy flow through communities. The ultimate disposition of the energy assimilated by consumers is by four routes: respiration, biomass accumulation, decay of organic matter by bacteria and other decomposer organisms, and consumption by consumers.

The actual amount of energy incorporated in the tissues of consumers at each tropic level is not determined by the gross amounts consumed. Rather it is the amount of organic energy converted into actual biomass. Consumers lose significant amounts of consumed energy due to assimilation inefficiencies, morphological and physiological maintenance, reproduction, and the process of finding and capturing food. The energy to perform the latter three processes is supplied by respiration.

Thus, the number of trophic levels that can be maintained in any ecosystem is finite. The limit is reached when consumers can no longer consume enough energy to balance the energy lost in the assimilation process, morphological and physiological maintanence, growth, reproduction, and in finding the food. Normally, ecosystems have about four or five trophic levels.

Finally, some ecosystems, like rivers and lakes, are characterized by large imports of biomass in the form of dead organic matter. This situation can cause consumer production to be higher than the autotrophs found in these systems.

Organic Decomposition and the Detritus Food Chain

The detritus food chain differs from the grazing food chain in several significant ways. First, the organisms making it up are in general physically smaller. Second, the functional roles of the different organisms do not fall as neatly into categories like the grazing food chain's trophic levels. Finally, detritivores live in environments (soil, sea bed, etc.) that are rich in scattered food particles. As a result, decomposers are less mobile than herbivores or carnivores.

The organisms of the detritus food chain include members of many different species of animals and plants, such as algae, bacteria, slime molds, fungi, protozoa, insects, mites, crustaceans, centipedes, mollusks, worms, sea cucumbers, and even some vertebrates. These organisms consume organic wastes, shed tissues, and the dead bodies of both plants and animals.

Earthworms are one of the most important soil decomposers. These organisms consume vast amounts of organic matter and mineral soil. As the organic matter passes through their digestive system, it is subjected to digestive enzymes and the grinding action of mineral soil particles. The amount of material consumed per day is often equal to their body weight.

Decomposers tend to always be active, processing large amounts of organic matter and releasing a great deal of energy mostly as heat from metabolic activities (e.g., compost heap). The end result of decomposition is the conversion of organic matter back into its original inorganic nutrient form. In mature forest and grassland soils, the decomposition process establishs an equilibrium over time where litterfall additions equal the amount of organic matter decomposed.

The rate of decomposition in a detritus food chain is controlled by many factors. In most terrestrial ecosystems, temperature and soil oxygen and moisture content tend to be the primary variables controlling the activities of decomposers. In some ecosystems, oxygen may not be readily available. In these cases aerobic respiration cannot take place, and the breakdown of organic compounds and energy extraction must then proceed by anaerobic means like fermentation. Organisms involved in fermentation include bacteria and yeast. Anaerobic decomposition releases much less energy from organic matter than does aerobic respiration. The end products of anaerobic decomposition include molecules such as carbon dioxide, water, and humus. They also include small-molecule alcohols, organic acids, ptomaines, amines, and other products, as well as gaseous substances such as methane.

Because respiration is so much more efficient at releasing the energy contained in organic molecules, the activity of the detritus food chain is much higher in an aerobic environment, and the breakdown of materials more complete. Organic matter breakdown is substantially slower and less complete in anaerobic environments. It also results in the accumulation of undegraded organic matter in the form of peat, organic soils, and highly organic sediments.

Trophic Pyramids and Food Webs

So far we have described food chains as morphological systems of energy flow. The energy flow within food chains can also be described in more quantitative terms. Several different quantitative models are commonly seen in the academic literature. One of these models, called a pyramid of biomass, quantifies all of the living biomass found in each of the trophic levels. Biomass can be defined as the weight of living matter (usually measured in dry weight per unit area).

In most ecosystems, the amount of biomass found in each trophic level decreases progressively as one moves from the beginning to the end of the grazing food chain. As described in previous sections, primary producers or plants are the original source of fixed organic energy in ecosystems. The energy that plants fix supports the life found in all other trophic levels. However, only the primary consumers or herbivores directly feed on the primary producers.

The amount of organic energy incorporated into the biomass of the primary consumers is significantly smaller than the amount found in the primary producer level. Theoretically, herbivores could consume all available plant life within an ecosystem. However, in reality this rarely occurs because plants have developed, through evolution, a number of mechanisms that protect most of their tissues from consumption. As a result, herbivores can only consume a portion of the available plant biomass. All that is consumed does not become herbivore biomass.

Significant losses of biomass occur because of digestive inefficiencies and respiration. Assimilation efficiencies for most terrestrial herbivores range from 20 to 60 percent. Some of the assimilated biomass is lost through the process of respiration.

The next level in the pyramid of biomass is the secondary consumers or primary carnivores. These organisms harvest a portion of the herbivore biomass for their nutrition. Once again, not all herbivores are eaten because of defensive mechanisms. Most herbivores possess some evolutionary adaptation that generally protects them from carnivore attack. These adaptations include the ability to fly and run, body armor, quills and protective spines, and camouflage. In general, carnivores have higher assimilation efficiencies than herbivores. Their assimilation efficiencies range from 50 to 90 percent. Only a portion of the assimilated organic energy becomes carnivore biomass because of the metabolic energy needs of body maintenance, growth, reproduction, and locomotion.

Many food chains have no more than four or five trophic levels. In the example above, the studied ecosystem had four. This final level is composed of the tertiary consumers who feed on the secondary consumers. The amount of biomass found in this trophic level is very small relative to the other levels.

Trophic pyramids have also been constructed to show the transfer of energy in caloric terms and the number of organisms found in each trophic level.

A model describing the organisms found in a food chain is called a food web. Food webs describe the complex patterns of energy flow in an ecosystem by modeling who consumes who. The illustration below describes a portion of the food web for a typical tidal marsh ecosystem located on the southern coast of British Columbia.

BIOGEOCHEMICAL CYCLING: INPUTS AND OUTPUTS OF NUTRIENTS TO ECOSYSTEMS

The patterns of cycling nutrients in the biosphere involves not only metabolism by living organisms, but also a series of strictly abiotic chemical reactions. Understanding the cycle of a single element requires the knowledge of a process that depends jointly on the biology of all organisms that utilize the element, its geological availability, and its organic and inorganic chemistry. Thus, understanding the cycling of biologically important elements is truly interdisciplinary in nature. We generally call this process biogeochemical cycling.

All biogeochemical cycles for all ecologically significant elements have both an organic and an inorganic component. Both are extremely important! How efficiently the nutrients move through the organic component back to the inorganic reserviors determines how much is available to organisms over the short term. The major reservoirs for all metabolically important elements are found either in the atmosphere, lithosphere (mainly rock, soil and other weathered sediments) or hydrosphere. Flow in the inorganic phase generally tends to be slower than in the organic phase.

Nutrients Essential for Life

Living organisms require the availability of about 20 to 30 chemical elements for the various of metabolic processes that take place in their bodies. Some products of this metabolism require relatively few nutrients for their production. For example, carbohydrates are photosynthesized from just water and carbon dioxide. Some organic substances, like amino acids and proteins, are more complex in their chemical make up and therefore require a number of different nutrients.

The types of nutrient needed by life is often categorized into two groups. Elements required in relatively large amounts are generally referred to as macronutrients. Macronutrients that constitute more than 1 % each of dry weight include carbon, oxygen, hydrogen, nitrogen, and phosphorus. Macronutrients that constitute 0.2 to 1 % of dry organic weight include sulfur, chlorine, potassium, sodium, calcium, magnesium, iron, and copper.

Nutrients needed in trace amounts are generally called micronutrients. These elements often constitute less than 0.2 % of dry organic matter. Some common micronutrients required by living organisms include aluminum, boron, bromine, chromium, cobalt, fluorine, gallium, iodine, manganese, molybdenum, selenium, silicon, strontium, tin, titanium, vanadium, and zinc.

Nutrient Inputs to Ecosystems

Important nutrients for life generally enter ecosystems by way of four processes:

Weathering

Rock weathering is one of the most important long-term sources for nutrients. However, this process adds nutrients to ecosystems in relatively small quantities over long periods of time. Important nutrients released by weathering include:

Calcium, magnesium, potassium, sodium, silicon, iron, aluminum, and phosphorus. Carbon, oxygen and nitrogen are not inputted into ecosystems by weathering. The main source for these important elements is the atmosphere and the decomposition of organic matter.

Atmospheric Input

Large quantities of nutrients are added to ecosystems from the atmosphere. This addition is done either through precipitation or by a number of biological processes.

— Carbon - absorbed by way of photosynthesis.

— Nitrogen - produced by lightning and precipitation.

Sulfur, chloride, calcium, and sodium - deposited by way of precipitation.

The quantity of nitrogen added to ecosystems by lightning and rain annually ranges from 1 to 20 kilograms per hectare depending on geographical location. A value of 5 to 8 kilograms per hectare is typical for temperate ecosystems like deciduous forest or grasslands.

Biological Nitrogen Fixation

Biological nitrogen fixation is the biochemical process where elemental nitrogen is combined into organic forms by metabolic processes. It is carried out by a limited number of organisms including several species of bacteria (not all are associated with legumes), a few actinomycetes (fungi), and blue-green algae (cyanobacteria).

The amount of nitrogen fixed by biological process has been estimated to be quite large. Recent calculations suggest that total annual global fixation of nitrogen is about 175 million metric tons.

Symbiotic Fixation with Legumes: The symbiosis of legumes (13,000 species) and bacteria of the genus Rhizobium provides the major biological source of fixed nitrogen in agricultural soils. Rhizobium bacteria invade the root hairs and the cortical cells, ultimately inducing the formation of nodules that serve as a home for the organisms. The host plant supplies the bacteria with carbohydrates for energy, and the bacteria reciprocate by supplying the plant with fixed nitrogen compounds, an association that is mutually beneficial.

The nitrogen fixed by Rhizobium goes in three directions. First, it is used directly by the host plant. Second, as the nitrogen passes from plant roots and nodules into the soil, some of it is mineralized and becomes available almost immediately as ammonium or nitrate compounds. The final pathway for the fixed nitrogen is that it can be incorporated into the bodies of the general purpose decay organisms and finally into the soil organic matter.

Symbiotic Fixation with Non-Legumes: About 160 species of non-legumes are known to develop nodules to house bacteria for nitrogen fixation. Some of the commonly known plant species that have this ability belong to the alder and birch family.

Symbiotic Nitrogen Fixation Without Nodules: Recent studies have called attention to several significant nonlegume symbiotic associations that do not produce plant nodules. Among the most important are those involving blue-green algae (cyanobacteria).

Immigration

The immigration of motile animals into an ecosystem can sometimes add significant additions of nutrients to an ecosystem that are locked up in the biomass of the organisms. These nutrients are released when the organism dies or sheds its tissues.

Nutrient Outputs to Ecosystems

Important nutrients required for life leave ecosystems by way of four processes:

Erosion

Soil erosion is probably the most import means of nutrient loss to ecosystems. Erosion is very active in agricultural and forestry systems, where cultivation, grazing, and clearcutting leaves the soil bare and unprotected. When unprotected, the surface of the soil is easily transported by wind and moving water. The top most layers of a soil, which have an abundance of nutrient rich organic matter, are the major storehouse for soil nutrients like phosphorus, potassium, and nitrogen.

Leaching

Another important process of nutrient loss is leaching. Leaching occurs when water flowing vertically through the soil transports nutrients in solution downward in the soil profile. Many of these nutrients can be completely lost from the soil profile if carried into groundwater and then horizontally transported into rivers, lakes, or oceans. Leaching losses are, generally, highest in disturbed ecosystems. In undisturbed ecosystems, efficient nutrient cycling limits the amount of nutrients available for this process.

Gaseous Losses

High losses of nutrients can also occur when specific environmental conditions promote the export of nutrients in a gaseous form. When the soil is wet and anaerobic, many compounds are chemically reduced to a gas from solid forms in the soil. This is especially true of soil nitrogen. Scientific studies in Netherlands have shown that about 80 % of the nitrogen fertilizer applied to the soil for crop consumption may be lost through the process of denitrification.

Emigration and Harvesting

Just as material may be introduced to ecosystems by migration, so too may it be lost. The emigration of animals, and the removal of vegetation by humans are both processes by which outputs can occur from an ecosystem.

Soil Organic Matter Decomposition and Nutrient Cycling

In terrestrial *ecosystems* most of the nutrient cycling occurs in the topmost horizons in *soil*. The main *inputs* to the soil come from *weathering*, rainfall, fertilizers, *atmospheric* sources and organisms. Under natural conditions, inputs from plants are the most important, including not only nutrients released by *organic decomposition*, but also substances washed in from the plant leaves (*foliar leaching*). Losses or *outputs* are by *leaching, erosion,* gaseous loss (like *denitrification*) and plant uptake. Within the soil, nutrients are stored on the soil particles, dead *organic matter,* or in chemical compounds.

Organic matter decomposition is the main process that recycles nutrients back into the soil. Decomposition of organic matter begins with large soil organisms like *earthworms, arthropods* (ants, beetles, and termites), and *gastropods* (slugs and snails). These organisms breakdown the organic matter into smaller pieces which can be decomposed by smaller organisms like *fungi* and heterotrophic *bacteria.*

Decomposition of organic matter may take several months to several years to complete. In tropical regions, the whole process is quite quick because moist conditions and high temperatures enhance biological activity.

The Carbon Cycle

All life is based on the element carbon. Carbon is the major chemical constituent of most organic matter, from fossil fuels to the complex molecules (DNA and RNA) that control genetic reproduction in organisms. Yet by weight, carbon is not one of the most abundant elements within the Earth's crust. In fact, the lithosphere is only 0.032 % carbon by weight. In comparison, oxygen and silicon respectively make up 45.2 % and 29.4 % of the Earth's surface rocks.

Carbon is stored on our planet in the following major sinks (Table 5): (1) as organic molecules in living and dead organisms found in the biosphere; (2) as the gas carbon dioxide in the atmosphere; (3) as organic matter in soils; (4)

in the lithosphere as fossil fuels and sedimentary rock deposits such as limestone, dolomite and chalk; and (5) in the oceans as dissolved atmospheric carbon dioxide and as calcium carbonate shells in marine organisms.

Table 5: Estimated major stores of carbon on the Earth

Sink	Amount in Billions of Metric Tons
Atmosphere	578 (as of 1700) - 766 (as of 1999)
Soil Organic Matter	1500 to 1600
Ocean	38,000 to 40,000
Marine Sediments and Sedimentary Rocks	66,000,000 to 100,000,000
Terrestrial Plants	540 to 610
Fossil Fuel Deposits	4000

Ecosystems gain most of their carbon dioxide from the atmosphere. A number of autotrophic organisms have specialized mechanisms that allow for absorption of this gas into their cells. With the addition of water and energy from solar radiation, these organisms use photosynthesis to chemically convert the carbon dioxide to carbon-based sugar molecules. These molecules can then be chemically modified by these organisms through the metabolic addition of other elements to produce more complex compounds like proteins, cellulose, and amino acids. Some of the organic matter produced in plants is passed down to heterotrophic animals through consumption.

Carbon dioxide enters the waters of the ocean by simple diffusion. Once dissolved in seawater, the carbon dioxide can remain as is or can be converted into carbonate (CO_3^{-2}) or bicarbonate (HCO_3^-). Certain forms of sea life biologically fix bicarbonate with calcium (Ca^{+2}) to produce calcium carbonate ($CaCO_3$). This substance is used to produce **shells** and other body parts by organisms such as coral, clams, oysters, some protozoa, and some algae. When these organisms die, their shells and body parts sink to the ocean floor where they accumulate as carbonate-rich deposits. After long periods of time, these deposits are physically and chemically altered intc sedimentary rocks. Ocean deposits are by far the biggest sink of carbon on the planet (Table 5).

Carbon is released from ecosystems as carbon dioxide gas by the process of respiration. Respiration takes place in both plants and animals and involves the breakdown of carbon-based organic molecules into carbon dioxide gas and some other compound by products. The detritus food chain contains a number of organisms whose primary ecological role is the decomposition of organic matter into its abiotic components.

Over the several billion years of geologic history, the quantity of carbon dioxide found in the atmosphere has been steadily decreasing. Researchers theorized that this change is in response to an increase in the sun's output over the same time period. Higher levels of carbon dioxide helped regulate the Earth's temperature to levels slightly higher than what is perceived today. These moderate temperatures allowed for the flourishing of plant life despite the lower output of solar radiation.

An enhanced greenhouse effect, due to the greater concentration of carbon dioxide gas in the atmosphere, supplemented the production of heat energy through higher levels of longwave counter-radiation. As the sun grew more intense, several biological mechanisms gradually locked some of the atmospheric carbon dioxide into fossil fuels and sedimentary rock. In summary, this regulating process has kept the Earth's global average temperature essentially constant over time. Some scientists suggest that this phenomena is proof for the Gaia hypothesis.

Carbon is stored in the lithosphere in both inorganic and organic forms. Inorganic deposits of carbon in the lithosphere include fossil fuels like coal, oil, and natural gas, oil shale, and carbonate based sedimentary deposits like limestone. Organic forms of carbon in the lithosphere include litter, organic matter, and humic substances found in soils. Some carbon dioxide is released from the interior of the lithosphere by volcanoes. Carbon dioxide released by volcanoes enters the lower lithosphere when carbon-rich sediments and sedimentary rocks are subducted and partially melted beneath tectonic boundary zones.

Since the Industrial Revolution, humans have greatly increased the quantity of carbon dioxide found in the Earth's atmosphere and oceans. Atmospheric levels have increased by over 30 %, from about 275 parts per million (ppm) in the early 1700s to just over 365 PPM today. Scientists estimate that future atmospheric levels of carbon dioxide could reach an amount between 450 to 600 PPM by the year 2100. The major sources of this gas due to human activities include fossil fuel combustion and the modification of natural plant cover found in grassland, woodland, and forested ecosystems.

Emissions from fossil fuel combustion account for about 65 % of the additional carbon dioxide currently found in the Earth's atmosphere. The other 35 % is derived from deforestation and the conversion of natural ecosystems into agricultural systems. Researchers have shown that natural ecosystems can store between 20 to 100 times more carbon dioxide than agricultural land-use types.

The Nitrogen Cycle

The nitrogen cycle represents one of the most important nutrient cycles found in terrestrial ecosystems. Nitrogen is used by living organisms to produce a number of complex organic molecules like amino acids, proteins, and nucleic acids. The largest store of nitrogen is found in the atmosphere where it exists as a gas (mainly N_2). The atmospheric store is about one million times larger than the total nitrogen contained in living organisms. Other major stores of nitrogen include organic matter in soil and the oceans.

Despite its abundance in the atmosphere, nitrogen is often the most limiting nutrient for plant growth. This problem occurs because most plants can only take up nitrogen in two solid forms: ammonium ion (NH^{4+}) and the ion nitrate (NO^{3-}). Most plants obtain the nitrogen they need as inorganic nitrate from the soil solution. Ammonium is used less by plants for uptake because in large concentrations it is extremely toxic. Animals receive the required nitrogen they need for metabolism, growth, and reproduction by the consumption of living or dead organic matter containing molecules composed partially of nitrogen.

In most ecosystems nitrogen is primarily stored in living and dead organic matter. This organic nitrogen is converted into inorganic forms when it re-enters the biogeochemical cycle via decomposition. Decomposers, found in the upper soil layer, chemically modify the nitrogen found in organic matter from ammonia (NH_3) to ammonium salts (NH^{4+}). This process is known as mineralization and it is carried out by a variety of bacteria, actinomycetes, and fungi.

Nitrogen in form of ammonium can be absorbed onto the surfaces of clay particles in the soil. The ion of ammonium has a positive molecular charge is normally held by soil colloids. This process is sometimes called micelle fixation. Ammonium is released from the colloids by way of cation exchange. When released, most of the ammonium is often chemically altered by a specific type of autotrophic bacteria (bacteria that belong to the genus Nitrosomonas) into nitrite (NO^{2-}). Further modification by another type of bacteria (belonging to the genus Nitrobacter) converts the nitrite to nitrate (NO^{3-}). Both of these processes involve chemical oxidation and are known as nitrification.

However, nitrate is very soluble and it is easily lost from the soil system by leaching. Some of this leached nitrate flows through the hydrologic system until it reaches the oceans where it can be returned to the atmosphere by denitrification. Denitrification is also common in anaerobic soils and is carried out by heterotrophic bacteria. The process of denitrification involves the reduction of nitrate into nitrogen (N_2) or nitrous oxide (N_2O) gas. Both of these gases then diffuse into the atmosphere. This process is important to the bacteria because it supplies them with oxygen for respiration.

Almost all of the nitrogen found in any terrestrial ecosystem originally came from the atmosphere. Small proportions enter the soil in rainfall or through the effects of lightning. The majority, however, is biochemically fixed within the soil by specialized micro-organisms like bacteria, actinomycetes, and cyanobacteria. Members of the bean family (legumes) and some other kinds of plants form mutualistic symbiotic relationships with nitrogen fixing bacterial.

In exchange for some nitrogen, the bacteria receive from the plants carbohydrates and special structures (nodules) in roots where they can exist in a moist environment. Scientist estimate that biological fixation globally adds approximately 140 million metric tons of nitogen to ecosystems every year. The activities of humans have severely altered the nitrogen cycle. Some of the major processes involved in this alteration include:

— The application of nitrogen fertilizers to crops has caused increased rates of denitrification and leaching of nitrate into groundwater. The additional nitrogen entering the groundwater system eventually flows into streams, rivers, lakes, and estuaries. In these systems, the added nitrogen can lead to eutrophication.

— Increased deposition of nitrogen from atmospheric sources because of fossil fuel combustion and forest burning. Both of these processes release a variety of solid forms of nitrogen through combustion.

7

Lithosphere

The Rock Cycle

The rock cycle is a general model that describes how various geological processes create, modify, and influence rocks. This model suggests that the origin of all rocks can be ultimately traced back to the solidification of molten magma. Magma consists of a partially melted mixture of elements and compounds commonly found in rocks. Magma exists just beneath the solid crust of the Earth in an interior zone known as the mantle.

Igneous rocks form from the cooling and crystallization of magma as it migrates closer to the Earth's surface. If the crystallization process occurs at the Earth's surface, the rocks created are called extrusive igneous rocks. Intrusive igneous rocks are rocks that form within the Earth's solid lithosphere. Intrusive igneous rocks can be brought to the surface of the Earth by denudation and by a variety of tectonic processes.

All rock types can be physically and chemically decomposed by a variety of surface processes collectively known as weathering. The debris that is created by weathering is often transported through the landscapc by erosional processes via streams, glaciers, wind, and gravity. When this debris is deposited as a permanent sediment, the processes of burial, compression, and chemical alteration can modify these materials over long periods of time to produce sedimentary rocks.

A number of geologic processes, like tectonic folding and faulting, can exert heat and pressure on both igneous and sedimentary rocks causing them to be altered physically or chemically. Rocks modified in this way are termed metamorphic rocks.

All of the rock types described above can be returned to the Earth's interior by tectonic forces at areas known as subduction zones. Once in the Earth's interior, extreme pressures and temperatures melt the rock back into magma to begin the rock cycle again.

Geologic Time

Geologists and geomorphologists describe the Earth's geologic history through a temporal system known as the geologic time scale. On this scale, time is measured using the following four units of time: eons, eras, periods and epochs. All of these temporal subdivisions are established on the occurrence of some important geologic event. For example, Hadean Eon represents the time on Earth when life did not exist. During the Archean Eon life started and was dominated by one-celled prokaryotic life forms. Eukaryotic one-celled organisms became dominant in the Proterozoic Eon. Multicellular organisms ruled the planet during the eon known as the Phanerozoic.

Concept of Uniformitarianism

Uniformitarianism is one of the most important unifying concepts in the geosciences. This concept developed in the late 1700s, suggests that catastrophic processes were not responsible for the landforms that existed on the Earth's surface. This idea was diametrically opposed to the ideas of that time period which were based on a biblical interpretation of the history of the Earth. Instead, the theory of uniformitarianism suggested that the landscape developed over long periods of time through a variety of slow geologic and geomorphic processes.

The term uniformitarianism was first used in 1832 by William Whewell, a University of Cambridge scholar, to present an alternative explanation for the origin of the Earth. The prevailing view at that time was that the Earth was created through supernatural means and had been affected by a series of catastrophic events such as the biblical Flood. This theory is called catastrophism.

The ideas behind uniformitarianism originated with the work of Scottish geologist James Hutton. In 1785, Hutton presented at the meetings of the Royal Society of Edinburgh that the Earth had a long history and that this history could be interpreted in terms of processes currently observed. For example, he suggested that deep soil profiles were formed by the weathering of bedrock over thousands of years. He also suggested that supernatural theories were not needed to explain the geologic history of the Earth.

Hutton's ideas did not gain major support of the scientific community until the work of Sir Charles Lyell. In the three volume publication Principles of Geology, Lyell presented a variety of geologic evidence from England, France, Italy, and Spain to prove Hutton's ideas correct and to reject the theory of catastrophism.

The theory of uniformitarianism was also important in shaping the development of ideas in other disciplines. The work of Charles Darwin and Alfred Wallace on the origin of the Earth's species extended the ideas of uniformitarianism into the biological sciences. The theory of evolution is based on the principle that the diversity seen in the Earth's species can be explained by the uniform modification of genetic traits over long periods of time.

Thus, uniformitarianism suggests that the continuing uniformity of existing processes should be used as the framework for understanding the geomorphic and geologic history of the Earth. Today, most theories of landscape evolution use the concept of uniformitarianism to describe how the various landforms of the Earth came to be.

Composition of Rocks

A rock can be defined as a solid substance that occurs naturally because of the effects of three basic geological processes: magma solidification; sedimentation of weathered rock debris; and metamorphism. As a result of these processes, three main types of rock occur:

— *Igneous Rocks* - produced by solidification of molten magma from the mantle. Magma that solidifies at the Earth's surface conceives extrusive or volcanic igneous rocks. When magma cools and solidifies beneath the surface of the Earth intrusive or plutonic igneous rocks are formed.

— *Sedimentary Rocks* - formed by burial, compression, and chemical modification of deposited weathered rock debris or sediments at the Earth's surface.

— *Metamorphic Rocks* - created when existing rock is chemically or physically modified by intense heat or pressure.

Most rocks are composed of minerals. Minerals are defined by geologists as naturally occurring inorganic solids that have a crystalline structure and a distinct chemical composition. Of course, the minerals found in the Earth's rocks are produced by a variety of different arrangements of chemical elements. A list of the eight most common elements making up the minerals found in the Earth's rocks is described in Table 1.

Table 1: Common elements found in the Earth's rocks

Element	Chemical Symbol	Percent Weight in Earth's Crust
Oxygen	O	46.60
Silicon	Si	27.72
Aluminum	Al	8.13
Iron	Fe	5.00
Calcium	Ca	3.63
Sodium	Na	2.83
Potassium	K	2.59
Magnesium	Mg	2.09

Over 2000 minerals have been identified by earth scientists. Table 2 describes some of the important minerals, their chemical composition, and classifies them in one of nine groups. The Elements Group includes over one hundred known minerals. Many of the minerals in this class are composed of only one element.

Geologists sometimes subdivide this group into metal and nonmetal categories. Gold, silver, and copper are examples of metals. The elements sulfur and carbon produce the minerals sulfur, diamonds, and graphite which are nonmetallic.

The Sulfide Group are an economically important class of minerals. Many of these minerals consist of metallic elements in chemical combination with the element sulfur. Most ores of important metals such as mercury (cinnabar - HgS), iron (pyrite - FeS_2), and lead (galena - PbS) are extracted from sulfides. Many of the sulfide minerals are recognized by their metallic luster.

The Halides are a group of minerals whose principle chemical constituents are fluorine, chlorine, iodine, and bromine. Many of them are very soluble in water. Halides also tend to have a highly ordered molecular structure and a high degree of symmetry. The most well-known mineral of this group is halite (NaCl) or rock salt.

The Oxides are a group of minerals that are compounds of one or more metallic elements combined with oxygen, water, or hydroxyl (OH). The minerals in this mineral group show the greatest variations of physical properties. Some are hard, others soft. Some have a metallic luster, some are clear and transparent. Some representative oxide minerals include corundum, cuprite, and hematite.

The Carbonates Group consists of minerals which contain one or more metallic elements chemically associated with the compound CO_4. Most carbonates are lightly colored and transparent when relatively pure. All carbonates are soft and brittle. Carbonates also effervesce when exposed to warm hydrochloric acid. Most geologists considered the Nitrates and Borates as subcategories of the carbonates. Some common carbonate minerals include calcite, dolomite, and malachite.

The Sulfates are a mineral group that contain one or more metallic element in combination with the sulfate compound SO4 . All sulfates are transparent to translucent and soft. Most are heavy and some are soluble in water. Rarer sulfates exist containing substitutions for the sulfate compound. For example, in the chromates SO4 is replaced by the compound CrO_4. Two common sulfates are anhydrite and gypsum.

The Phosphates are a group of minerals of one or more metallic elements chemically associated with the phosphate compound PO_4. The phosphates are often classified together with the arsenate, vanadate, tungstate, and molybdate minerals. One common phosphate mineral is apatite. Most phosphates are heavy but soft. They are usually brittle and occur in small crystals or compact aggregates.

The Silicates are by far the largest group of minerals. Chemically, these minerals contain varying amounts of silicon and oxygen. It is easy to distinguish silicate minerals from other groups, but difficult to identify individual minerals within this group. None are completely opaque. Most are light in weight. The construction component of all silicates is the tetrahedron. A tetrahedon is a chemical structure where a silicon atom is joined by four oxygen atoms (SiO_4). Some representative minerals include albite, augite, beryl, biotite, hornblende, microcline, muscovite, olivine, othoclase, and quartz.

The Organic minerals are a rare group of minerals chemically containing hydrocarbons. Most geologists do not classify these substances as true minerals. Note that our original definition of a mineral excludes organic substances. However, some organic substances that are found naturally on the Earth that exist as crystals that resemble and act like true minerals. These substances are called organic minerals. Amber is a good example of an organic mineral.

Table 2: Classification of some of the important minerals found in rocks.

Group (and information link)	Typical Minerals	Chemistry
Elements	Gold	Au
	Silver	Ag
	Copper	Cu
	Carbon (Diamond and Graphite)	C
	Sulfur	S
Sulfides	Cinnabar	HgS
	Galena	PBS
	Pyrite	FeS_2
Halides	Fluorite	CaF_2
	Halite	NaCl
Oxides	Corundum	Al_2O_3
	Cuprite	Cu_2O
	Hematite	Fe_2O_3
Carbonates (Nitrates and Borates)	Calcite	$CaCO_3$
	Dolomite	$CaMg(CO_3)_2$
	Malachite	$Cu_2(CO_3)(OH)_2$
Sulfates	Anhydrite	$CaSO_4$
	Gypsum	$CaSO_4$-2(H_2O)
Phosphates (Arsenates, Vanadates, Tungstates, and Molybdates)	Apatite	$Ca_5(F,Cl,OH)(PO_4)$
Silicates	Albite	$NaAlSi_3O_8$
	Augite	$(Ca, Na)(Mg, Fe, Al)(Al, Si)_2O_6$
	Beryl	$Be_3Al_2(SiO_3)_6$
	Biotite	$K(FE,Mg)_3AlSi_3O_{10}(F, OH)_2$
	Hornblende	$Ca_2(Mg, Fe, Al)_5(Al,Si)_8$

		$O_{22}(OH)_2$
		Microcline
$KAlSi_3O_8$		
	Muscovite	$KAl_2(AlSi_3O_{10})(F, OH)_2$
	Olivine	$(Mg, Fe)_2SiO_4$
	Orthoclase	$KAlSi_3O_8$
	Quartz	SiO_2
	Organics	Amber

Characteristics of Igneous Rocks

Chemistry of the rock material, pressure, presence of gases (like water vapor) all influence when melting occurs. Most of the heat required to melt rock into magma comes from the Earth's central internal region known as the core. Scientists estimate that the temperature of the Earth's core is about 5000° Celsius. Heat moves from the Earth's core towards the solid outer crust by convection and conduction.

Convection moves hot plumes of magma vertically from the lower mantle to the upper mantle. Some of these plumes melt through the Earth's solid lithosphere and can produce intrusive igneous features and extrusive igneous features on the surface. Heat can also be generated in the lower lithosphere through friction. The tectonic movement of subducted crustal plates can generate enough heat (and pressure) to melt rock. This fact explains the presence of volcanoes along the margin of some continental plates.

Types of Igneous Rocks

The type of igneous rocks that form from magma is a function of three factors: the chemical composition of the magma; temperature of solidification; and the rate of cooling which influences the crystallization process. Magma can vary chemically in its composition. For example, the amount of silica (SiO_2) found in magma can vary from 75 % to less than 45 %. The temperature of cooling determines which types of minerals are found dominating the rock's composition. Rocks that begin their cooling at low temperatures tend to be rich in minerals composed of silicon, potassium, and aluminum.

High temperature igneous rocks are dominated by minerals with higher quantities of calcium, sodium, iron, and magnesium. The rate of cooling is important in crystal development. Igneous rocks that form through a gradual cooling process tend to have large crystals. Relatively fast cooling of magma produces small crystals. Volcanic magma that cools very quickly on the Earth's surface can produce obsidian glass which contains no crystalline structures.

Geologists have classified the chemistry of igneous rocks into four basic types: felsic, intermediate, mafic, and ultramafic. Igneous rocks derived from felsic magma contain relatively high quantities of sodium, aluminum, and potassium and are composed of more than 65 % silica. Rocks formed from felsic magma include granite, granodiorite, dacite, and rhyolite. All of these rock types are light in color because of the dominance of quartz, potassium and sodium feldspars, and plagioclase feldspar minerals.

Dacite and granodiorite contain slightly more biotite and amphibole minerals than granite and rhyolite. Rhyolite and dacite are produced from continental lava flows that solidify quickly. The quick solidification causes the mineral crystals in these rocks to be fine grained. Granite and granodiorite are common intrusive igneous rocks that are restricted to the Earth's continents. Large expanses these rocks were formed during episodes of mountain building on the Earth. Because granite and granodiorite form beneath the Earth's surface their solidification is a relatively slow process. This slow solidification produces a rock with a coarse mineral grain.

Mafic magma produces igneous rocks rich in calcium, iron, and magnesium and are relatively poor in silica (silica amounts from 45 to 52 %). Some common mafic igneous rocks include fine grained basalt and coarse grained gabbro. Mafic igneous rocks tend to be dark in color because they contain a large proportion of minerals rich in iron and magnesium (pyroxene, amphiboles, and olivine). Basalt is much more common than gabbro. It is found in the upper portion of the oceanic crust and also in vast continental lava flows that cover parts of Washington, Oregon, Idaho, and California. Gabbro is normally found in the lower parts of oceanic crust and sometimes in relatively small intrusive features in continental crust.

Andesite and diorite are intermediate igneous rocks that have a chemistry between mafic and felsic (silica amounts between 53 to 65 %). These rocks are composed predominantly of the minerals plagioclase feldspar, amphibole, and pyroxene. Andesite is a common fine grained extrusive igneous rock that forms from lavas erupted by volcanoes located along continental margins. Coarse grained diorite is found in intrusive igneous bodies associated with continental crust.

Ultramafic igneous rocks contain relative low amounts of silica (< 45 %) and are dominated by the minerals olivine, calcium-rich plagioclase feldspars, and pyroxene. Peridotite is the most common ultramafic rock found in the Earth's crust. These rocks are extremely rare at the Earth's surface.

Igneous Rocks and the Bowen Reaction Series

In the 1920s, N.L. Bowen theorized that the formation of minerals, which make up igneous rocks, begins with two different chemical sequences at high temperatures that eventually merge into a single series at cooler temperatures. One sequence, the discontinuous series, involves the formation of chemically unique minerals at discrete temperature intervals from iron and magnesium rich mafic magma. In the other sequence, known as the continuous series, temperature reduction causes a gradual change in the chemistry of the minerals that form calcium and sodium rich felsic magma.

The discontinuous series starts with the formation of rocks that are primarily composed of the mineral olivine. Continued temperature decreases change the minerals dominating the composition of the rock from pyroxene, to amphibole, and then biotite. The continuous series produces light colored rocks rich in plagioclase feldspar minerals. At high temperatures, the plagioclase feldspar minerals are dominated with the element calcium. With continued cooling, the calcium in these minerals is gradually replaced with sodium.

The convergence of both series occurs with a continued drop in magma temperature. In the merged series, the minerals within the crystallizing rock become richer in potassium and silica and we get the formation of first potassium feldspars and then the mineral muscovite. The last mineral to crystallize in the Bowen reaction series is quartz. Quartz is a silicate mineral composed of just silicon and oxygen (SiO_2).

Characteristics of Sedimentary Rocks

Sedimentary rocks can be categorized into three groups based on sediment type. Most sedimentary rocks are formed by the lithification of weathered rock debris that has been physically transported and deposited. During the transport process, the particles that make up these rocks often become rounded due to abrasion or can become highly sorted. Examples of this type of sedimentary rock include conglomerate and sandstone. Scientists sometimes call this general group of sedimentary rocks clastic. The remaining types of sedimentary rocks are created either from chemical precipitation and crystallization, or by the lithification of once living organic matter. We identify these sedimentary rocks as being non-clastic.

All sedimentary rocks are lithified into some collective mass. Lithification is any process that turns raw rock sediment into consolidated sedimentary rock. The process of lithification usually produces identifiable layering in these type of rocks. Lithification can occur by way of:

— Drying and compaction.

— Oxidation of iron and aluminum.

— Precipitation of calcium and silica.

The classification of clastic sedimentary rocks is based on the particle types found in the rock. Some types of clastic sedimentary rocks are composed of weathered rock material like gravel, sand, silt, and clay. Others can be constructed from the break up and deposition of shells, coral and other marine organisms by wave-action and ocean currents. There were two types of non-clastic sedimentary rocks. One group forms through the chemical precipitation and crystallization of elements and compounds from solution. Elements such as calcium, sodium, potassium, and magnesium are commonly released into the environment through a variety of chemical weathering processes. These elements can then become dissolved into aqueous solutions that are often transported via runoff, stream flow, or groundwater flow. If this solution enters a basin environment where evaporation exceeds precipitation and in-flow, sedimentary evaporites can form because of the loss of water from the solution.

The oceans are almost saturated with dissolved calcium carbonate. This compound originates from the shells of a variety of marine organisms that use it for the construction of shells and other hard body parts. Because these organisms are surrounded in a solution, some of the calcium carbonate dissolves into the ocean waters. Under the right circumstances the dissolved calcium carbonate can precipitate out forming chemically created limestone deposits. The formation of dolomite involves the chemical modification of limestone deposits by a magnesium rich solution.

The following table describes some of the common forms of chemical precipitated sedimentary rocks.

Table 3: Sedimentary rocks formed as chemical precipitates

Name of Rock	Precipitate Type
Halite	Sodium and Chlorine
Gypsum	Calcium, Sulfur, and Oxygen
Silcretes	Silica
Ferricretes	Iron
Limestone	Calcium Carbonate
Dolomite	Calcium Magnesium Carbonate

Several types of sedimentary rocks are formed from the lithification of once living organisms. Limestone deposits can be formed by the direct lithification of coral reefs, marine organism shells, or marine organism skeletons. Chalk is a particular variety of limestone that is composed of the skeletons of marine microorganisms like forminifera. Coal and lignite are the lithified remains of plants.

Characteristics of Metamorphic Rocks

Metamorphism involves the alteration of existing rocks by either excessive heat and pressure, or through the chemical action of fluids. This alteration can cause chemical changes or structural modification to the minerals making up the rock. Structural modification may involve the simple reorganization of minerals into layers or the aggregation of minerals into specific areas within the rock.

Much of the Earth's continental crust is composed of metamorphic and igneous rocks. Together, these two rock types form the base material at the core of the Earth's major continental masses. Overlying this core are often thick layers of sedimentary rocks. In some regions, this base rock is exposed to the atmosphere and is known as shields. Metamorphic rocks also are the rock type found at the core of the world's various mountain ranges.

Heat and Metamorphism

Heat is an important agent in the metamorphic modification of rock. Rocks begin to change chemically at temperatures above 200° Celsius. At these temperatures, the crystalline structure of the minerals in the rock are broken down and transformed using different combinations of the available elements and compounds. As a result, new minerals are created.

The metamorphic process stops when the temperatures become high enough (600 to 1200° Celsius) to cause complete melting of the rock. If rocks are heated to the point where they become magma, the magma when cooled creates new igneous rocks. Thus, metamorphism only refers to the alteration of rock that takes place before complete melting occurs.

Heat can be applied to rock through two processes: tectonic subduction and the intrusion of magma. Some rocks that are formed at the surface are subsequently transported deep into the crust and the upper mantle at tectonic subduction zones. Temperatures beneath the Earth's surface increase with depth at a rate of about 25° Celsius per kilometer. Scientists estimate that the temperature at the base of the crust is about 800 to 1200° Celsius. This heat is generated from the decay of radioactive materials in the mantle and the core.

Magma can sometimes migrate up through the crust forming an igneous intrusion. This is especially true along continental boundaries, like the western side of North America, where subduction is taking place. Metamorphism takes place in the rock surrounding the magma body because of heat dissipation. Because of the nature of the dissipation process, the level of metamorphic alteration in the influenced rock decreases with distance from the igneous intrusion.

Pressure and Metamorphism

Rocks that buried are subjected to pressure because of the weight of overlying materials. Pressure can also be exerted on rocks due the forces involved in a variety of tectonic processes. The most obvious effect of pressure on rocks is the reorientation of mineral crystals. Under extreme levels of pressure rocks become plastic creating flow structures in their crystalline structure. Pressure almost never acts in isolation as temperatures do get higher with increasing depth below the Earth's surface.

Chemical Action of Fluids

Water and carbon dioxide are often found in small amounts in the perimeter between mineral crystals or in the pore spaces of rocks. When mixed, the resulting fluid enhances metamorphism by dissolving ions and by causing chemical reactions. Usually, the end product of this process is the creation of new minerals by the substitution, removal, or addition of chemical ions. Sometimes fluids can also permeate into rock from adjacent magma.

Types of Metamorphism

Metamorphism can occur by way of the following three processes.

— Thermal metamorphism involves the heating and structural and chemical alteration of rocks through processes associated with plate tectonics. This type of metamorphism has two sub-categories:

— Regional metamorphism is the large scale heating and modification of existing rock through the creation of plutons at tectonic zones of subduction. It involves large areas and large volumes of rock.

— Contact metamorphism is the small scale heating and alteration of rock by way of a localized igneous intrusion (for example, volcanic dykes or sills).

— Dynamic metamorphism causes only the structural alteration of rock through pressure. The minerals in the altered rocks do not change chemically. The extreme pressures associated with mountain building can cause this type of metamorphism.

— Metasomatic metamorphism involves the chemical replacement of elements in rock minerals when gases and liquids permeate into bedrock.

Common Metamorphic Rocks

Examples of metamorphic rock types include: Slate is a fine grained metamorphic rock. It is created by minor metamorphism of shale or mudstone. This rock is characterized by the foliation of its mineral grains which causes it to have cleavage that is parallel.

Schist is a medium to coarse grained foliated rock. Foliation is the result of the rearrangement of mica, chlorite, talc, and hematite mineral grains into parallel structures. When compared to slate, schists result from more intense metamorphism.

Gneiss is a coarse grained metamorphized igneous rock. In this rock, you get the recrystallization and foliation of quartz, feldspars, micas, and amphiboles into alternating light and dark colored bands.

Structure of the Earth

The Earth is an oblate spheroid. It is composed of a number of different layers as determined by deep drilling and seismic evidence. These layers are:

— The core which is approximately 7000 kilometers in diameter (3500 kilometers in radius) and is located at the Earth's center.

— The mantle which surrounds the core and has a thickness of 2900 kilometers.

— The crust floats on top of the mantle. It is composed of basalt rich oceanic crust and granitic rich continental crust.

The core is a layer rich in iron and nickel that is composed of two layers: the inner and outer cores. The inner core is theorized to be solid with a density of about 13 grams per cubic centimeter and a radius of about 1220 kilometers. The outer core is liquid and has a density of about 11 grams per cubic centimeter. It surrounds the inner core and has an average thickness of about 2250 kilometers.

The mantle is almost 2900 kilometers thick and comprises about 83 % of the Earth's volume. It is composed of several different layers. The upper mantle exists from the base of the crust downward to a depth of about 670 kilometers. This region of the Earth's interior is thought to be composed of peridotite, an ultramafic rock made up of the minerals olivine and pyroxene.

The top layer of the upper mantle, 100 to 200 kilometers below surface, is called the asthenosphere. Scientific studies suggest that this layer has physical properties that are different from the rest of the upper mantle. The rocks in this upper portion of the mantle are more rigid and brittle because of cooler temperatures and lower pressures. Below the upper mantle is the lower mantle that extends from 670 to 2900 kilometers below the Earth's surface. This layer is hot and plastic. The higher pressure in this layer causes the formation of minerals that are different from those of the upper mantle.

The lithosphere is a layer that includes the crust and the upper most portion of the asthenosphere. This layer is about

100 kilometers thick and has the ability to glide over the rest of the upper mantle. Because of increasing temperature and pressure, deeper portions of the lithosphere are capable of plastic flow over geologic time. The lithosphere is also the zone of earthquakes, mountain building, volcanoes, and continental drift.

The topmost part of the lithosphere consists of crust. This material is cool, rigid, and brittle. Two types of crust can be identified: oceanic crust and continental crust. Both of these types of crust are less dense than the rock found in the underlying upper mantle layer. Ocean crust is thin and measures between 5 to 10 kilometers thick. It is also composed of basalt and has a density of about 3.0 grams per cubic centimeter.

The continental crust is 20 to 70 kilometers thick and composed mainly of lighter granite. The density of continental crust is about 2.7 grams per cubic centimeter. It is thinnest in areas like the Rift Valleys of East Africa and in an area known as the Basin and Range Province in the western United States. Continental crust is thinkest beneath mountain ranges and extends into the mantle. Both of these crust types are composed of numerous tectonic plates that float on top of the mantle. Convection currents within the mantle cause these plates to move slowly across the asthenosphere.

The lithosphere consists of the oceanic crust, continental crust, and uppermost mantle. Beneath the lithosphere is the asthenosphere. This layer, which is also part of the upper mantle, extends to a depth of about 200 kilometers. Sedimentary deposits are commonly found at the boundaries between the continental and oceanic crust.

One interesting property of the continental and oceanic crust is that these tectonic plates have the ability to rise and sink. This phenomenon, known as isostacy, occurs because the crust floats on top of the mantle like ice cubes in water. When the Earth's crust gains weight due to mountain building or glaciation, it deforms and sinks deeper into the mantle. If the weight is removed, the crust becomes more buoyant and floats higher in the mantle.

This process explains recent changes in the height of sea-level in coastal areas of eastern and northern Canada and Scandinavia. Some locations in these regions of the world have seen sea-level rise by as much as one meter over the last one hundred years. This rise is caused by isostatic rebound. Both of these areas where covered by massive glacial ice sheets about 10,000 years ago. The weight of the ice sheets pushed the crust deeper into the mantle. Now that the ice is gone, these areas are slowly increasing in height to some new equilibrium level.

Plate Tectonics

In the 19th and early 20th centuries, several scientists suggested that the continental masses had the ability to move across the Earth's surface. These early theories of continental drift were based on the following evidence:

- Locations of fossil occurrences suggested that some of the continental masses may have been connected in the geological past.
- Paleoclimatic evidence indicates that now tropical regions on some continents had polar climates in the past. This may indicate that these regions were located at different latitudes.
- Some geologic deposits of rocks on the East coast of North and South America are similar to deposits found on the West coast of Africa and Europe.

During the first 30 years of this century the theory of continental drift was actively debated among geo-scientists. However, during the following 30 year period, debate on this theory waned because of the inability of scientists to propose a mechanism to cause the movement of the continental masses.

In the 1960s, the theory was resurrected with the discovery of alternating patterns of rock magnetism in surface sea-floor rocks. Scientists had previously discovered that the magnetic orientation of certain crystals in rocks varies from normal to reversed polarity depending on the date that the rock was formed and solidified. It was also discovered that these magnetic reversals were common and occurred on a regular basis.

The polarity patterns found in the rocks at the ocean floor seemed to mirror themselves either side of the mid-oceanic ridge found at the centers of the ocean basins. Further, geologic dating of the rocks indicated the age of the sea-floor rocks increase as one moved away from the mid-oceanic ridge. Based on this information, scientists developed the theory of sea-floor spreading which suggested that volcanic rift zones at the mid-oceanic ridge represent areas of crustal creation.

The theory of sea-floor spreading started a revolution in the Earth Sciences. Subsequent research discovered that the Earth's surface was composed of a number of oceanic and continental plates that float on top of the asthenosphere. Other research suggested that convection currents within the Earth's mantle were responsible for the creation of oceanic crust and the drifting of the continents.

The theory of plate tectonics offered new and more scientifically sound explanations for a number of observed geologic phenomena. In this type of a collision, one of the plates is subducted under the other creating a deep oceanic trench. The Marianas trench in the Pacific ocean is created by the collision of the fast-moving Pacific Plate against the slower moving Philippine Plate. Convergence of two oceanic plates also creates chains of volcanic islands called island arcs. Island arcs are created by the friction of subduction which creates hot plumes of magma at the interface of the two plates. These hot plumes of magma then rise to the Earth's surface to form volcanoes. Another phenomena associated with collision and subduction of the plates is earthquakes. The gliding of one plate under the other is not smooth but jerky producing seismic waves.

Crustal Formation Processes

Studies of seismic waves have discovered that the Earth's crust consists of two basic types. Beneath the oceans we find a crust that is on average 7 kilometers thick and composed mainly of basalt. Oceanic crust also has a density of about 3.0 grams per cubic centimeter.

The continent's are composed of mainly granitic rock whose thickness varies between 10 and 70 kilometers. The thickest portions of continental crust are found under the various mountain ranges. Continental crust is also lighter than oceanic crust having a density of about 2.7 grams per cubic centimeter. Oceanic and continental rocks also differ from each other in terms of age. Continental crust contains some very old rocks that were formed during the Precambrian between 3 and 4 billion years ago. Oceanic rocks are normally quite young deposits. Isotopic dating of the rocks found on the sea-floor indicates that they were created less that 180 million years ago.

Variations in the age, density, and chemical composition of oceanic and continental crust suggest that these lithospheric deposits were created by different processes.

Continental Crust

All of the Earth's continents have a core foundation that is made of mixtures of very old granite, gneiss, schist, sedimentary, and volcanic rocks. This core foundation is often referred to as a shield or basement rock. Rocks found in the shields were formed during the Precambrian and are some of the oldest rocks found on the Earth. In Canada, some of the metamorphic rocks have been dated to an age of 3.96 billion years. Geologists believe that the major continental cores were formed by the early solidification of the lighter components of magma between 3.9 and 3.8 billion years ago.

The continental shields are generally covered by younger sedimentary deposits. These sedimentary rocks constitute the interior platforms of the continents. The oldest platform rocks were laid down in shallow seas about 600 million years ago. In central North America, the platform sedimentary deposits are between 1000 to 2000 meters thick. Together the shield and platform form what geologists call a craton. Most of the Earth's continental cratons have been tectonically stable and have experienced little deformation for hundreds of millions of years.

Around the edge of the continental cratons are the continental margins. The continental margins are primarily composed of sedimentary rocks. These sedimentary rocks were originally laid down in the oceans. Tectonic collisions and plate subduction caused the accretion of these deposits at the edges of the continental cratons. In some cases, this accretion is modified by the processes of tectonic compression, folding, and faulting to produce mountain ranges.

The movement of oceanic crust under the continents is the mechanism responsible for transporting the sedimentary materials found on the ocean floor to the edge of the continents. The animation also describes how friction between the oceanic and continental plates creates hot plumes of magma which migrate up through the continental mass to form plutonic and extrusive igneous features.

Igneous Activity and the Continents

Materials are also added to continental crust through intrusive and extrusive igneous activity. Plumes of magma from the Earth's asthenosphere are generated from the friction produced at the contact zone where oceanic crust slides past continental crust. These plumes then rise upward into continental crust to form granitic plutons or a variety of volcanic features on the Earth's surface. A pluton can be defined as any igneous intrusion of rock that forms a kilometer or more below the Earth's surface.

Some of the features associated with igneous intrusions or plutons. Some of the major features include:

— *Dyke*: thin vertical veins of igneous rock that form in the fractures found within the crust. Because these intrusive features cool quickly their rocks are dominated with fine mineral grains.

— *Sill*: horizontal planes of solidified magma that run parallel to the grain of the original rock deposit.

— *Batholith*: large plutonic masses of intrusive rock with more than 100 square kilometers of surface area.

— *Volcanic Pipe*: if a dyke reaches the surface of the Earth it is then called a volcanic pipe. Igneous deposits produced by this feature are extrusive in nature.

Plumes that are able to reach the Earth's surface produce volcanoes. Most of the continental volcanoes found on our planet are located along the edge of the continents where oceanic crust is being actively subducted. In North America, the zone of active volcanoes is located on the west coast where the subduction of the Pacific plate occurs. Volcanoes add mass to the continents when magma produces lava flows, tephra, and volcanic ash.

Oceanic Crust

Unlike continental crust, oceanic crust is actively being created at the various mid-oceanic ridges. At the mid-ocean ridges, magma erupts onto the ocean floor in centrally located rift zones. The newly added rock then horizontally pushes previously created ocean crust away from the rift in a conveyor belt fashion. When the oceanic crust encounters a slab of continental crust it becomes subducted because of its greater density. This process causes the oceanic crust to return to the mantle were it is re-melted into magma. The process also causes the movement of continental crust across the surface of the Earth.

Formation of oceanic crust at rift zones located in the mid-oceanic ridges. Beneath the rift zone upwellings of magma occur in the mantle. These upwellings produce fissures and volcanoes on the ocean floor surface. The added rock, produced from the solidification of magma, pushes previously formed oceanic crust horizontally away from the rift zone like a conveyor belt. Ocean crust is returned to the mantle through subduction. This can occur when ocean crust meets continental crust or other ocean plates

Mountain Building

A mountain can be defined as an area of land that rises abruptly from the surrounding region. A mountain range is a succession of many closely spaced mountains covering a particular region of the Earth. Mountain belts consist of several mountain ranges that run roughly parallel to each other. The North American Cordillera, the Himalayas, the Alps, and the Appalachians are all examples of mountain belts that are composed of numerous mountain ranges.

Some mountains are volcanic in origin forming where rising magma breaks through the Earth's surface. Volcanic mountains tend to have sporadic distributions within a mountain range (Mount St. Helens, Rainier, and Baker) or can occur alone because of a localized hot spot (Hawaiian Islands). Most mountains were created from tectonic forces that elevate, fold, and fault rock materials. Tectonic mountains can occur as a single range (the Urals) or as a belt of several mountain ranges (North American Cordillera).

These major mountain systems include the North American Cordillera, Andes, Alps, Urals, Appalachians, Himalaya, Caledonian Belt, and the Tasman Belt. Some of the major mountain belts on the Earth are the North American Cordillera (A), Appalachians (B), Caledonian Belt (C), Andes (D), Urals (E), Himalaya (F), Alps (G), and the Tasman Belt (H).

The Earth's mountain ranges have various ages of formation. Parts of the Himalayas are relatively quite young.

Mountain building in this region of the world began about 45 million years ago when the continental plates of India and Eurasia converged on each other. The Himalaya mountains are still actively being uplifted. The Appalachian belt is quite old. Mountain building in this region of the world started about 450 million years ago. Orogeny stopped in the Appalachians about 250 million years ago. The long passage of time without active uplift has allowed weathering and erosion to remove large amounts of bedrock from the Appalachians. These processes have also significantly lowered and rounded the peaks of the various mountains found in this belt.

Mountain building episodes in the North American Cordillera have been occurring over a very long period of time and still continue today. Some sedimentary rocks in the Rocky Mountain range (located on the eastern edge of the North American Cordillera) date to over a billion years old.

Evolution of Mountains

Geologists have developed a general model to explain how most mountain ranges form. This model suggests that mountain building involves three stages:

(1) accumulation of sediments,

(2) an orogenic period of rock deformation and crustal uplift, and

(3) a period of crustal uplift caused by isostatic rebound and block-faulting.

Mountain belts normally contain numerous layers of sedimentary and volcanic igneous rocks. These accumulations can be several kilometers in thickness. Most of these accumulations were originally deposited in a marine environment. The beds of the sedimentary rocks are composed of particles that came from nearby terrestrial landmasses. These particles were released from rocks by weathering and then transported by erosional forces to the edge of the terrestrial continental crust.

Beyond the edge of the continents, these sediments are lithified to form shales, limestones, and sandstones that make up the continental shelves, slopes, and rise. Accumulations of volcanic rock develop along convergent boundaries where subduction is causing magma plumes to form plutons and volcanoes. The volcanoes are usually spatially organized in a line, called an island arc, that runs at right angles to the direction of crustal movement.

The subduction of one oceanic plate under another creates friction that melts rock into magma. This magma then migrates upward through the crust forming plutons and volcanoes. Volcanic rocks can also be converted into sedimentary sandstones. Weathering and erosion can remove material from terrestrial volcanic deposits to marine depositional environments.

In the orogenic stage of mountain building, the accumulated sediments become deformed by compressional forces from the collision of tectonic plates. This tectonic convergence can be of three types: arc-continent, ocean-continent or continent-continent. In an ocean-continent convergence, the collision of ocean and continental plates causes the accretion of marine sedimentary deposits to the edge of the continent. Arc-continent convergence occurs when an island arc collides with the edge of a continental plate.

In this convergence, the ocean plate area between the arc and the continent is subducted into the asthenosphere and the volcanic rocks and sediments associated with the island arc become accreted to the margin of the continent over time. This type of collision may have been responsible for the creation of the Sierra Nevada mountains in California during the Mesozoic Era. The final type of convergence occurs when an ocean basin closes and two continental plates collide. Continent-continent convergence mountain building is responsible for the formation of the Himalayas, Ural, and Appalachian mountain systems.

In all three types of tectonic convergence, layered rocks that were once located in the ocean basin are squeezed into a smaller and smaller area. This compression causes the once flat sedimentary beds to be folded and uplifted. When the compressional forces become greater than the rocks ability to deform, faulting occurs. Compressional forces typically result in reverse and overthrust faulting. Another consequence of the orogenic stage is regional metamorphism and the incursion of magma plumes, plutons, and volcanoes into the growing mountain range.

At the end of plate convergence, mountain building enters its final stage. This stage is characterized by crustal uplift because of isostatic rebound and block-faulting. Isostatic rebound involves the vertical movement of continental crust that is floating in the plastic upper mantle. As erosion removes surface materials from mountains, the weight of the crust in this region becomes progressively less. With less weight, the continental crust makes an isostatic adjustment causing it to rise vertically (float higher) in the mantle. This process also causes tensional forces to exist in a horizontal direction breaking the continental crust into a number of blocks. Each block moves vertically to compensate for the tensional forces producing normal and graben faults.

Crustal Deformation Processes: Folding and Faulting

Deformation of rock involves changes in the shape and/or volume of these substances. Changes in shape and volume occur when stress and strain causes rock to buckle and fracture or crumple into folds. A fold can be defined as a bend in rock that is the response to compressional forces. Folds are most visible in rocks that contain layering. For plastic deformation of rock to occur a number of conditions must be met, including:

— The rock material must have the ability to deform under pressure and heat.

— The higher the temperature of the rock the more plastic it becomes.

— Pressure must not exceed the internal strength of the rock. If it does, fracturing occurs.

— Deformation must be applied slowly.

A number of different folds have been recognized and classified by geologists. The simplest type of fold is called a monocline. This fold involves a slight bend in otherwise parallel layers of rock.

Faults form in rocks when the stresses overcome the internal strength of the rock resulting in a fracture. A fault can be defined as the displacement of once connected blocks of rock along a fault plane. This can occur in any direction with the blocks moving away from each other. Faults occur from both tensional and compressional forces.

There are several different kinds of faults. These faults are named according to the type of stress that acts on the rock and by the nature of the movement of the rock blocks either side of the fault plane. Normal faults occur when tensional forces act in opposite directions and cause one slab of the rock to be displaced up and the other slab down.

8

Maps and Remote Sensing

A map can be simply defined as a graphic representation of the real world. This representation is always an abstraction of reality. Because of the infinite nature of our Universe it is impossible to capture all of the complexity found in the real world. For example, topographic maps abstract the three-dimensional real world at a reduced scale on a two-dimensional plane of paper.

Maps are used to display both cultural and physical features of the environment. Standard topographic maps show a variety of information including roads, land-use classification, elevation, rivers and other water bodies, political boundaries, and the identification of houses and other types of buildings. Some maps are created with very specific goals in mind.

The art of map construction is called cartography. People who work in this field of knowledge are called cartographers. The construction and use of maps has a long history. Some academics believe that the earliest maps date back to the fifth or sixth century BC. Even in these early maps, the main goal of this tool was to communicate information. Early maps were quite subjective in their presentation of spatial information. Maps became more objective with the dawn of Western science. The application of scientific method into cartography made maps more ordered and accurate. Today, the art of map making is quite a sophisticated science employing methods from cartography, engineering, computer science, mathematics, and psychology.

Cartographers classify maps into two broad categories: reference maps and thematic maps. Reference maps normally show natural and human-made objects from the geographical environment with an emphasis on location. Examples of general reference maps include maps found in atlases and topographic maps. Thematic maps are used to display the geographical distribution of one phenomenon or the spatial associations that occur between a number of phenomena.

Map Projection

The shape of the Earth's surface can be described as an ellipsoid. An ellipsoid is a three-dimensional shape that departs slightly from a purely spherical form. The Earth takes this form because rotation causes the region near the equator to bulge outward to space. The angular motion caused by the Earth spinning on its axis also forces the polar regions on the globe to be somewhat flattened.

Representing the true shape of the Earth's surface on a map creates some problems, especially when this depiction is illustrated on a two-dimensional surface. To overcome these problems, cartographers have developed a number of standardized transformation processes for the creation of two-dimensional maps. All of these transformation processes create some type of distortion artifact. The nature of this distortion is related to how the transformation process modifies specific geographic properties of the map. Some of the geographic properties affected by projection distortion include: distance; area; straight line direction between points on the Earth; and the bearing of cardinal points from locations on our planet.

The illustrations below show some of the common map projections used today. The first two-dimensional projection shows the Earth's surface as viewed from space (Figure 1). This orthographic projection distorts distance, shape, and the size of areas. Another serious limitation of this projection is that only a portion of the Earth's surface can be viewed at any one time.

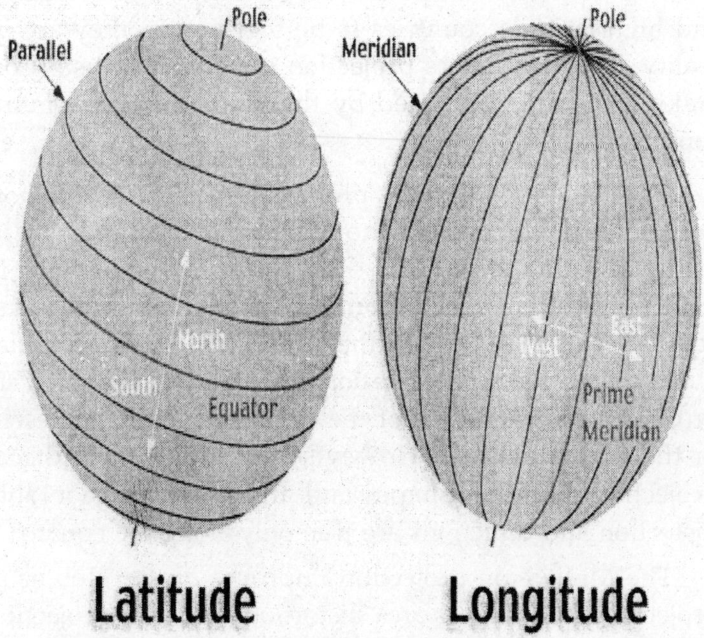

Figure 1: Earth as observed from a vantage point in space.

The second illustration displays a Mercator projection of the Earth. On a Mercator projection, the north-south scale increases from the equator at the same rate as the corresponding east-west scale. As a result of this feature, angles drawn on this type of map are correct. Distortion on a Mercator map increases at an increasing rate as one moves toward higher latitudes. Mercator maps are used in navigation because a line drawn between two points of the Earth has true direction. However, this line may not represent the shortest distance between these points.

The Gall-Peters projection was developed to correct some of the distortion found in the Mercator system. The Mercator projection causes area to be gradually distorted from the equator to the poles. This distortion makes middle and high latitude countries to be bigger than they are in reality. The Gall-Peters projection corrects this distortion making the area occupied by the world's nations more comparable.

The Miller Cylindrical projection is another common two-dimensional map used to represent the entire Earth in a rectangular area. In this project, the Earth is mathematically projected onto a cylinder tangent at the equator. This projection in then unrolled to produce a flat two-dimensional representation of the Earth's surface. This projection reduces some of the scale exaggeration present in the Mercator map. However, the Miller Cylindrical projection describes shapes and areas with considerable distortion and directions are true only along the equator.

The Mollweide projection improves on the Robinson projection and has less area distortion. The final projection presented presents areas on a map that are proportional to the same areas on the actual surface of the Earth. However, this Sinusoidal Equal-Area projection suffers from distance, shape, and direction distortions.

MAP SCALE

Maps are rarely drawn at the same scale as the real world. Most maps are made at a scale that is much smaller than the area of the actual surface being depicted. The amount of reduction that has taken place is normally identified somewhere on the map. This measurement is commonly referred to as the map scale. Conceptually, we can think of map scale as the ratio between the distance between any two points on the map compared to the actual ground distance represented. This concept can also be expressed mathematically as:

On most maps, the map scale is represented by a simple fraction or ratio. This type of description of a map's scale is called a representative fraction. For example, a map where one unit (centimeter, meter, inch, kilometer, etc.) on the illustration represents 1, 000, 000 of these same units on the actual surface of the Earth would have a representative fraction of 1/1, 000, 000 (fraction) or 1:1, 000, 000 (ratio). Of these mathematical representations of scale, the ratio form is most commonly found on maps.

Scale can also be described on a map by a verbal statement. For example, 1:1, 000, 000 could be verbally described as "1 centimeter on the map equals 10 kilometers on the Earth's surface" or "1 inch represents approximately 16 miles".

Most maps also use graphic scale to describe the distance relationships between the map and the real world. In a graphic scale, an illustration is used to depict distances on the map in common units of measurement. Graphic scales are quite useful because they can be used to measure distances on a map quickly.

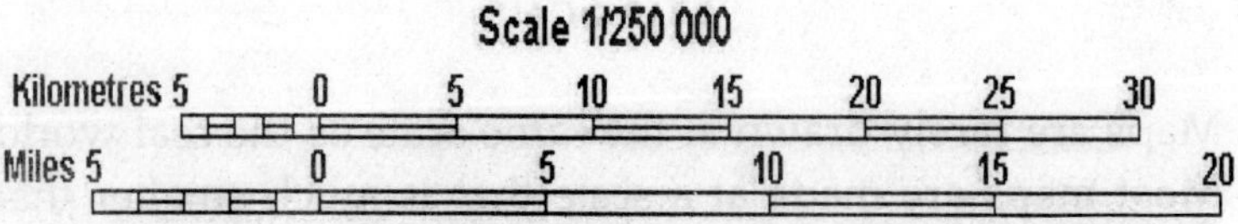

Figure 2: The following graphic scale was drawn for map with a scale of 1:250, 000. In the illustration distances in miles and kilometers are graphically shown.

Maps are often described, in a relative sense, as being either small scale or large scale.

Location, Distance, and Direction on Maps

Location on Maps

Most maps allow us to specify the location of points on the Earth's surface using a coordinate system. For a two-dimensional map, this coordinate system can use simple geometric relationships between the perpendicular axes on a grid system to define spatial location. Two types of coordinate systems are currently in general use in geography: the geographical coordinate system and the rectangular (also called Cartesian) coordinate system.

Geographical Coordinate System

The geographical coordinate system measures location from only two values, despite the fact that the locations are described for a three-dimensional surface. The two values used to define location are both measured relative to the polar axis of the Earth. The two measures used in the geographic coordinate system are called latitude and longitude.

Latitude measures the north-south position of locations on the Earth's surface relative to a point found at the centre of the Earth. This central point is also located on the Earth's rotational or polar axis. The equator is the starting point for the measurement of latitude. The equator has a value of zero degrees. A line of latitude or parallel of 30° North has an angle that is 30° north of the plane represented by the equator. The maximum value that latitude can attain is either 90° North or South. These lines of latitude run parallel to the rotational axis of the Earth.

Longitude measures the west-east position of locations on the Earth's surface relative to a circular arc called the Prime Meridian. The position of the Prime Meridian was determined by international agreement to be in-line with the location of the former astronomical observatory at Greenwich, England. Because the Earth's circumference is similar to circle, it was decided to measure longitude in degrees. The number of degrees found in a circle is 360. The Prime Meridian has a value of zero degrees.

A line of longitude or meridian of 45° West has an angle that is 45° west of the plane represented by the Prime Meridian. The maximum value that a meridian of longitude can have is 180° which is the distance halfway around a circle. This meridian is called the International Date Line. Designations of west and east are used to distinguish where a location is found relative to the Prime Meridian. For example, all of the locations in North America have a longitude that is designated west.

Universal Transverse Mercator System (UTM)

Another commonly used method to describe location on the Earth is the Universal Transverse Mercator (UTM) grid system. This rectangular coordinate system is metric, incorporating the meter as its basic unit of measurement. UTM also uses the Transverse Mercator projection system to model the Earth's spherical surface onto a two-dimensional plane. The UTM system divides the world's surface into 60 - six degree longitude wide zones that run north-south. These zones start at the International Date Line and are successively numbered in an eastward direction . Each zone stretches from 84° North to 80° South.

In the centre of each of these zones is a central meridian. Location is measured in these zones from a false origin which is determined relative to the intersection of the equator and the central meridian for each zone. For locations in the Northern Hemisphere, the false origin is 500, 000 meters west of the central meridian on the equator.

Coordinate measurements of location in the Northern Hemisphere using the UTM system are made relative to this point in meters in eastings (longitudinal distance) and northings (latitudinal distance). The point defined by the intersection of 50° North and 9° West would have a UTM coordinate of Zone 29, 500000 meters east (E), 5538630 meters north (N). In the Southern Hemisphere, the origin is 10, 000, 000 meters south and 500, 000 meters west of the equator and central meridian, respectively. The location found at 50° South and 9° West would have a UTM coordinate of Zone 29, 500000 meters E, 4461369 meters N .

The UTM system has been modified to make measurements less confusing. In this modification, the six degree wide zones are divided into smaller pieces or quadrilaterals that are eight degrees of latitude tall. Each of these rows is labeled, starting at 80° South, with the letters C to X consecutively with I and O being omitted. The last row X differs from the other rows and extends from 72 to

84° North latitude (twelve degrees tall). Each of the quadrilaterals or grid zones are identified by their number/ letter designation. In total, 1200 quadrilaterals are defined in the UTM system.

The quadrilateral system allows us to further define location using the UTM system. For the location 50° North and 9° West, the UTM coordinate can now be expressed as Grid Zone 29U, 500000 meters E, 5538630 meters N.

Each UTM quadrilateral is further subdivided into a number of 100, 000 by 100, 000 meter zones. These subdivisions are coded by a system of letter combinations where the same two-letter combination is not repeated within 18 degrees of latitude and longitude. Within each of the 100, 000 meter squares one can specify location to one-meter accuracy using a 5 digit eastings and northings reference system.

The UTM grid system is displayed on all United States Geological Survey (USGS) and National Topographic Series (NTS) of Canada maps. On USGS 7. 5-minute quadrangle maps (1:24, 000 scale), 15-minute quadrangle maps (1:50, 000, 1:62, 500, and standard-edition 1:63, 360 scales), and Canadian 1:50, 000 maps the UTM grid lines are drawn at intervals of 1, 000 meters. On USGS maps at 1:100, 000 and 1:250, 000 scale and Canadian 1:250, 000 scale maps a full UTM grid is shown at intervals of 10, 000 meters.

Distance on Maps

On a map that has a large scale, 1:125, 000 or larger, distance distortion is usually insignificant. An example of a large-scale map is a standard topographic map. On these maps measuring straight line distance is simple. Distance is first measured on the map using a ruler. This measurement is then converted into a real world distance using the map's scale. For example, if we measured a distance of 10 centimeters on a map that had a scale of 1:10, 000, we would multiply 10 (distance) by 10, 000 (scale). Thus, the actual distance in the real world would be 100, 000 centimeters.

Measuring distance along map features that are not straight is a little more difficult. One technique that can be employed for this task is to use a number of straight-line segments. The accuracy of this method is dependent on the number of straight-line segments used. Another method for measuring curvilinear map distances is to use a mechanical device called an opisometer. This device uses a small rotating wheel that records the distance travelled. The recorded distance is measured by this device either in centimeters or inches.

Direction on Maps

Like distance, direction is difficult to measure on maps because of the distortion produced by projection systems. However, this distortion is quite small on maps with scales larger than 1:125, 000. Direction is usually measured relative to the location of North or South Pole. Directions determined from these locations are said to be relative to True North or True South. The magnetic poles can also be used to measure direction. However, these points on the Earth are located in spatially different spots from the geographic North and South Pole.

The North Magnetic Pole is located at 78.3° North, 104.0° West near Ellef Ringnes Island, Canada. In the Southern Hemisphere, the South Magnetic Pole is located in Commonwealth Day, Antarctica and has a geographical location of 65° South, 139° East. The magnetic poles are also not fixed overtime and shift their spatial position overtime.

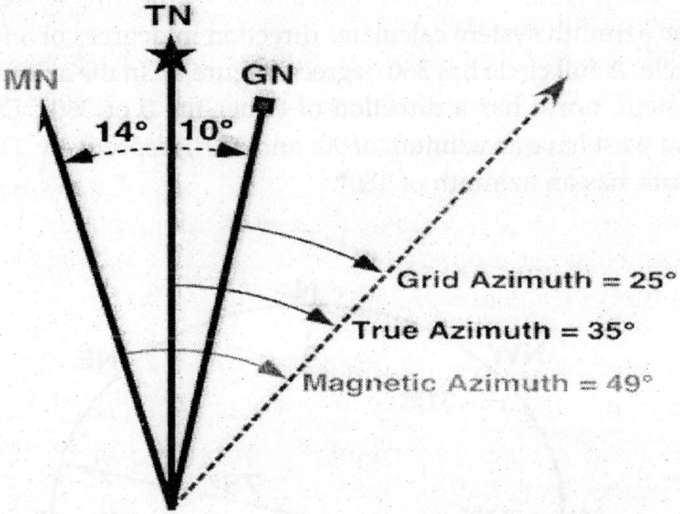

Figure 3: This declination diagram describes the angular difference between Grid, True, and Magnetic North.

Topographic maps normally have a declination diagram drawn on them (Figure 3). On Northern Hemisphere maps, declination diagrams describe the angular difference between Magnetic North and True North. On the map, the angle of True North is parallel to the depicted lines of longitude. Declination diagrams also show the direction of Grid North. Grid North is an angle that is parallel to the easting lines found on the Universal Transverse Mercator (UTM) grid system.

In the field, the direction of features is often determined by a magnetic compass which measures angles relative to Magnetic North. Using the declination diagram found on a map, individuals can convert their field measures of magnetic direction into directions that are relative to either

Grid or True North. Compass directions can be described by using either the azimuth system or the bearing system. The azimuth system calculates direction in degrees of a full circle. A full circle has 360 degrees (Figure 4). In the azimuth system, north has a direction of either the 0 or 360°. East and west have an azimuth of 90° and 270°, respectively. Due south has an azimuth of 180°.

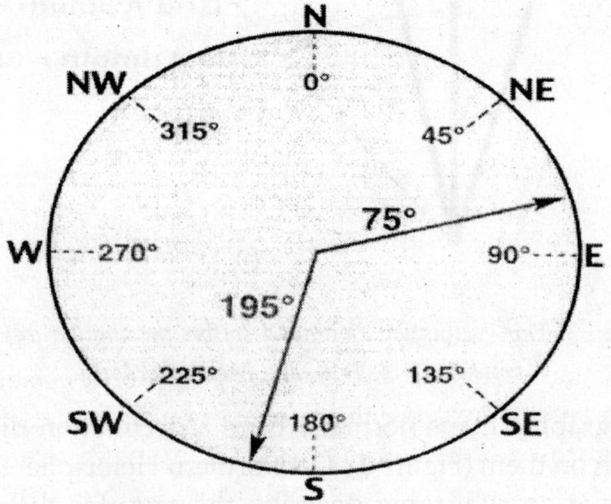

Figure 4: Azimuth system for measuring direction is based on the 360 degrees found in a full circle.

The bearing system divides direction into four quadrants of 90 degrees. In this system, north and south are the dominant directions. Measurements are determined in degrees from one of these directions. The measurement of two angles based on this system are described in Figure 5.

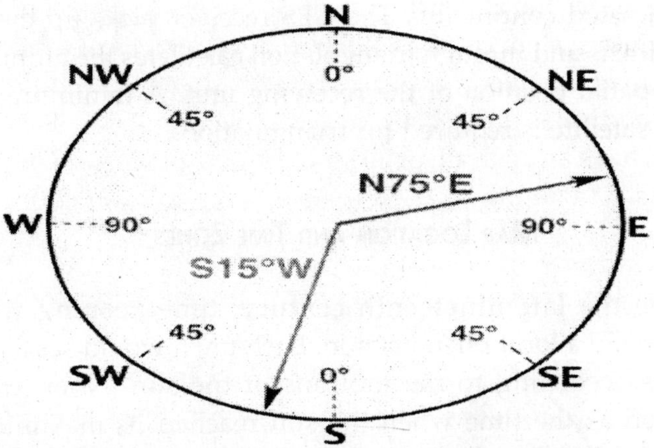

Figure 5: The bearing system uses four quadrants of 90 degrees to measure direction.

The illustration shows two direction measurements. These measurements are made relative to either north or south. North and south are given the measurement 0 degrees. East and west have a value of 90 degrees. The first measurement is found in the north - east quadrant. As a result, its measurement is north 75 degrees to the east or N75°E. The first measurement is found in the south - west quadrant. Its measurement is south 15 degrees to the west or S15°W.

Global Positioning Systems

Determination of location in field conditions was once a difficult task. In most cases, it required the use of a topographic map and landscape features to estimate location. However, technology has now made this task very simple. Global Positioning Systems (GPS) can calculate one's location to an accuracy of about 30-meters. These systems consist of two parts: a GPS receiver and a network of many satellites. Radio transmissions from the satellites are broadcasted continually. The GPS receiver picks up these broadcasts and through triangulation calculates the altitude and spatial position of the receiving unit. A minimum of three satellite is required for triangulation.

Map Location and Time Zones

Before the late nineteenth century, time keeping was essentially a local phenomenon. Each town would set their clocks according to the motions of the sun. Noon was defined as the time when the sun reached its maximum altitude above the horizon. Cities and towns would assign a clockmaker to calibrate a town clock to these solar motions. This town clock would then represent "official" time and the citizens would set their watches and clocks accordingly.

The later half of the nineteenth century was time of increased movement of humans. In the United States and Canada, large numbers of people were moving west and settlements in these areas began expanding rapidly. To support these new settlements, railroads moved people and resources between the various cities and towns. However, because of the nature of how local time was kept, the railroads experience major problems in constructing timetables for the various stops. Timetables could only become more efficient if the towns and cities adopted some type of standard method of keeping time.

In 1878, Canadian Sir Sanford Fleming suggested a system of worldwide time zones that would simplify the keeping of time across the Earth. Fleming proposed that the globe be divided into 24 time zones, each 15 degrees of longitude in width. Since the world rotates once every 24 hours on its axis and there are 360 degrees of longitude, each hour of Earth rotation represents 15 degrees of longitude.

Railroad companies in Canada and the United States began using Fleming's time zones in 1883. In 1884, an International Prime Meridian Conference was held in Washington D. C. to adopted the standardize method of time keeping and determined the location of the Prime Meridian. Conference members agreed that the longitude of Greenwich, England would become zero degrees longitude and established the 24 time zones relative to the Prime Meridian. It was also proposed that the measurement of time on the Earth would be made relative to the astronomical measurements at the Royal Observatory at Greenwich. This time standard was called Greenwich Mean Time (GMT).

Today, many nations operate on variations of the time zones suggested by Sir Fleming. In this system, time in the various zones is measured relative the Coordinated Universal Time (UTC) standard at the Prime Meridian. Coordinated Universal Time became the standard legal reference of time all over the world in 1972. UTC is determined from six primary atomic clocks that are coordinated by the International Bureau of Weights and Measures (BIPM) located in France.

Topographic Maps

A topographic map is a detailed and accurate two-dimensional representation of natural and human-made features on the Earth's surface. These maps are used for a number of applications, from camping, hunting, fishing, and hiking to urban planning, resource management, and surveying. The most distinctive characteristic of a topographic map is that the three-dimensional shape of the Earth's surface is modeled by the use of contour lines. Contours are imaginary lines that connect locations of similar elevation. Contours make it possible to represent the height of mountains and steepness of slopes on a two-dimensional map surface. Topographic maps also use a variety of symbols to describe both natural and human made features such as roads, buildings, quarries, lakes, streams, and vegetation.

Topographic maps produced by the Canadian National Topographic System (NTS) are generally available in two different scales: 1:50, 000 and 1:250, 000. Maps with a scale of 1:50, 000 are relatively large-scale covering an area approximately 1000 square kilometers. At this scale, features as small as a single home can be shown. The smaller scale 1:250, 000 topographic map is more of a general purpose reconnaissance-type map. A map of this scale covers the same area of land as sixteen 1:50, 000 scale maps.

In the United States, topographic maps have been made by the United States Geological Survey (USGS) since 1879. Topographic coverage of the United States is available at scales of 1:24, 000, 1:25, 000 (metric), 1:62, 250, 1:63, 360 (Alaska only), 1:100, 000 and 1:250, 000.

Topographic Map Symbols

Topographic maps use symbols to represent natural and human constructed features found in the environment. The symbols used to represent features can be of three types: points, lines, and polygons. Points are used to depict features like bridges and buildings. Lines are used to graphically illustrate features that are linear. Some common linear features include roads, railways, and rivers. However, we also need to include representations of area, in the case of forested land or cleared land; this is done through the use of color.

The set of symbols used on Canadian National Topographic System (NTS) maps has been standardized to simplify the map construction process. A description of the complete set of symbols available can be found in a published guide titled: Standards and Specifications for Polychrome Maps. This guide guarantees uniform illustration of surface features on both 1:50 000 and 1:250 000 topographic maps. Despite the existence of this guide, we can find that some topographic maps may use different symbols to depict a feature. This occurs because the symbols used are graphically refined over time - as a result the Standards and Specifications for Polychrome Maps guide is always under revision.

Contour Lines

Topographic maps can describe vertical information through the use of contour lines (contours). A contour line is an isoline that connects points on a map that have the same elevation. Contours are often drawn on a map at a uniform vertical distance. This distance is called the contour interval. Note that every fifth brown contour lines is drawn bold and has the appropriate elevation labeled on it. These contours are called index contours. The interval at which contours are drawn on a map depends on the amount of the relief depicted and the scale of the map.

Contour lines provide us with a simple effective system for describing landscape configuration on a two-dimensional map. The arrangement, spacing, and shape of the contours provide the user of the map with some idea of what the actual topographic configuration of the land surface looks like. Contour intervals the are spaced closely together describe a steep slope. Gentle slopes are indicated by widely spaced contours. Contour lines that V upwards indicate the presence of a river valley. Ridges are shown by contours that V downwards.

Topographic Profiles

A topographic profile is a two-dimensional diagram that describes the landscape in vertical cross-section. Topographic profiles are often created from the contour information found on topographic maps. The simplest way to construct a topographic profile is to place a sheet of blank paper along a horizontal transect of interest. From the map, the elevation of the various contours is transferred on to the edge of the paper from one end of the transect to the other. Now on a sheet of graph paper use the x-axis to represent the horizontal distance covered by the transect. The y-axis is used to represent the vertical dimension and measures the change in elevation along the transect.

Most people exaggerate the measure of elevation on the y-axis to make changes in relief stand out. Place the beginning of the transect as copied on the piece of paper at the intersect of the x and y-axis on the graph paper. The contour information on the paper's edge is now copied onto the piece of graph paper.

Military Grid Reference System and Map Location

Two rectangular grid systems are available on topographic maps for identifying the location of points. These systems are the Universal Transverse Mercator (UTM) grid system and the Military Grid Reference System. The Military Grid

Reference System is a simplified form of Universal Transverse Mercator grid system and it provides a very quick and easy method of referencing a location on a topographic map.

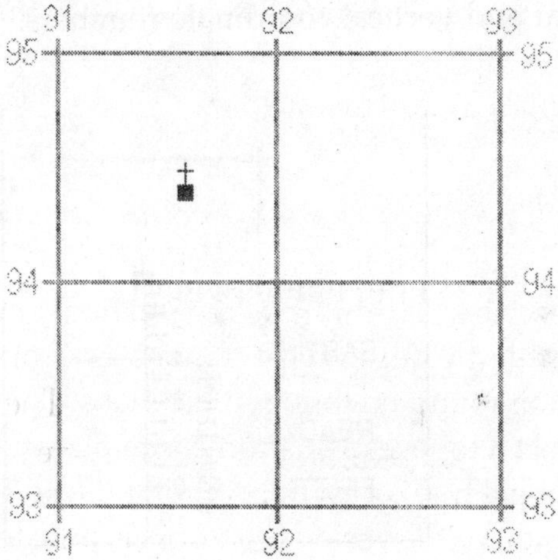

Figure 6: Portion of a Military Grid Reference System found on a topographic map. Coordinates on this system are based on a X (horizontal increasing from left to right) and Y (vertical increasing from bottom to top) system. The symbol depicting a church is located in the square 919

Identifying numbers for each of these horizontal and vertical lines are found along the map's margin. Each identifying number consists of two digits which range from a value of 00 to 99 (Figure 6). Each individual square in the grid system represents a distance of a 1000 by 1000 meters and the total size of the grid is 100, 000 by 100, 000 meters.

One problem associated with the Military Grid Reference System is the fact that reference numbers must be repeat every 100, 000 meters. To overcome this difficulty, a method was devised to identify each 100, 000 by 100, 000 meter grid with two identifying letters which are printed in on the border of all topographic maps. When making reference to a location with the Military Grid Reference System identifying letters are always given before the horizontal and vertical coordinate numbers.

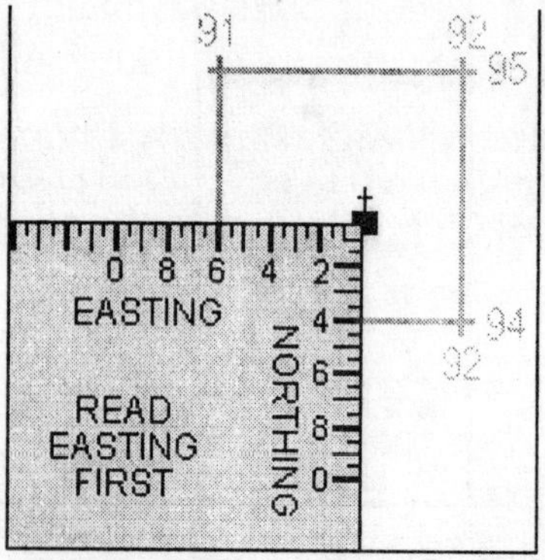

Figure 7: Further determination of the location of the church described in Figure 6. Using the calibrated ruler we can now suggest the location of the church to be 916 on the X-axis and 944 on the Y-axis. Note that the location reference always has an even number of digits, with the three digits representing the easting and the second three the northing.

Each individual square in the Military Grid Reference System can be further divided into 100 smaller squares (ten by ten). This division allows us to calculate the location of an object to within 100 meters. Figure 7 indicates that the church is six tenths of the way between lines 91 and 92, and four tenths of the way between lines 94 and 95. Using these values, we can state that the easting as being 916 and the northing as 944. By convention, these two numbers are combined into a coordinate reference of 916944.

Introduction to Remote Sensing

Remote sensing can be defined as the collection of data about an object from a distance. Humans and many other types of animals accomplish this task with aid of eyes or by the sense of smell or hearing. Geographers use the technique of remote sensing to monitor or measure phenomena found in the Earth's lithosphere, biosphere, hydrosphere, and atmosphere. Remote sensing of the environment by geographers is usually done with the help of mechanical devices known as remote sensors. These gadgets have a greatly improved ability to receive and record information about an object without any physical contact.

Often, these sensors are positioned away from the object of interest by using helicopters, planes, and satellites. Most sensing devices record information about an object by measuring an object's transmission of electromagnetic energy from reflecting and radiating surfaces. Remote sensing imagery has many applications in mapping land-use and cover, agriculture, soils mapping, forestry, city planning, archaeological investigations, military observation, and geomorphological surveying, among other uses. For example, foresters use aerial photographs for preparing forest cover maps, locating possible access roads, and measuring quantities of trees harvested.

Specialized photography using color infrared film has also been used to detect disease and insect damage in forest trees. The simplest form of remote sensing uses photographic cameras to record information from visible or near infrared wavelengths. In the late 1800s, cameras were positioned above the Earth's surface in balloons or kites to take oblique aerial photographs of the landscape.

During World War I, aerial photography played an important role in gathering information about the position and movements of enemy troops. These photographs were often taken from airplanes. After the war, civilian use of aerial photography from airplanes began with the systematic vertical imaging of large areas of Canada, the United States, and Europe. Many of these images were used to construct topographic and other types of reference maps of the natural and human-made features found on the Earth's surface.

The development of color photography following World War II gave a more natural depiction of surface objects. Color aerial photography also greatly increased the amount of information gathered from an object. The human eye can differentiate many more shades of color than tones of gray. In 1942, Kodak developed color infrared film, which recorded wavelengths in the near-infrared part of the electromagnetic spectrum. This film type had good haze penetration and the ability to determine the type and health of vegetation.

Satellite Remote Sensing

In the 1960s, a revolution in remote sensing technology began with the deployment of space satellites. From their high vantage-point, satellites have a greatly extended view of the Earth's surface. The first meteorological satellite, TIROS-1, was launched by the United States using an Atlas rocket on April 1, 1960. This early weather satellite used vidicon cameras to scan wide areas of the Earth's surface.

Early satellite remote sensors did not use conventional film to produce their images. Instead, the sensors digitally capture the images using a device similar to a television camera. Once captured, this data is then transmitted electronically to receiving stations found on the Earth's surface.

Today, the GOES (Geostationary Operational Environmental Satellite) system of satellites provides most of the remotely sensed weather information for North America. To cover the complete continent and adjacent oceans two satellites are employed in a geostationary orbit. The western half of North America and the eastern Pacific Ocean is monitored by GEOS-10, which is directly above the equator and 135° West longitude. The eastern half of North America and the western Atlantic are cover by GEOS-8.

The GEOS-8 satellite is located overhead of the equator and 75° West longitude. Advanced sensors aboard the GOES satellite produce a continuous data stream so images can be viewed at any instance. The imaging sensor produces visible and infrared images of the Earth's terrestrial surface and oceans. Infrared images can depict weather conditions even during the night. Another sounding sensor aboard the satellite provides data to determine vertical temperature profiles, vertical moisture profiles, total precipitable water, and atmospheric stability.

In the 1970s, the second revolution in remote sensing technology began with the deployment of the Landsat satellites. Since this 1972, several generations of Landsat satellites with their Multispectral Scanners (MSS) have been providing continuous coverage of the Earth for almost 30 years. Current, Landsat satellites orbit the Earth's surface at an altitude of approximately 700 kilometers. Spatial resolution of objects on the ground surface is 79 x 56 meters. Complete coverage of the globe requires 233 orbits and occurs every 16 days. The Multispectral Scanner records a zone of the Earth's surface that is 185 kilometers wide in four wavelength bands: band 4 at 0.5 to 0.6 micrometers, band 5 at 0.6 to 0.7 micrometers, band 6 at 0.7 to 0.8 micrometers, and band 7 at 0.8 to 1.1 micrometers.

A second sensing system was added to Landsat satellites launched after 1982. This imaging system, known as the Thematic Mapper, records seven wavelength bands from the visible to far-infrared portions of the electromagnetic spectrum. In addition, the ground resolution of this sensor was enhanced to 30 x 20 meters. This modification allows for greatly improved clarity of imaged objects.

The usefulness of satellites for remote sensing has resulted in several other organizations launching their own devices. In France, the SPOT (Centre National d'Etudes Spatiales) satellite program has launched four satellites since 1986. Since 1986, SPOT satellites have produced more than 5. 5 million images. SPOT satellites use two different sensing systems to image the planet. One sensing system produces black and white panchromatic images from the visible band (0. 51-0. 73 micrometers) with a ground resolution of 10 x 10 meters. SPOT-5 will be launched in late 2001. This satellite is much improved from its siblings sporting a ground resolution as low as 2.5 x 2.5 meters in panchromatic mode and 10 x 10 meters in multispectral operation.

Radarsat-1 was launched by the Canadian Space Agency in November, 1995. As a remote sensing device, Radarsat is quite different from the Landsat and SPOT satellites. Radarsat is an active remote sensing system that transmits and receives microwave radiation. Landsat and SPOT sensors passively measure reflected radiation at wavelengths roughly equivalent to those detected by our eyes. Radarsat's microwave energy penetrates clouds, rain, dust, or haze and produces images regardless of the sun's illumination allowing it to image in darkness. Radarsat images have a resolution between 8 to 100 meters. This sensor has found important applications in crop monitoring, defence surveillance, disaster monitoring, geologic resource mapping, sea-ice mapping and monitoring, oil slick detection, and digital elevation modeling.

Principles of Object Identification

Most people have no problem identifying objects from photographs taken from an oblique angle. Such views are natural to the human eye and are part of our everyday experience. However, most remotely sensed images are taken from an overhead or vertical perspective and from distances quite removed from ground level. Both of these circumstances make the interpretation of natural and human-made objects somewhat difficult. In addition, images obtained from devices that receive and capture electromagnetic wavelengths outside human vision can present views that are quite unfamiliar.

To overcome the potential difficulties involved in image recognition, professional image interpreters use a number of characteristics to help them identify remotely sensed objects. Some of these characteristics include:

— *Shape*: this characteristic alone may serve to identify many objects. Examples include the long linear lines of highways, the intersecting runways of an airfield, the perfectly rectangular shape of buildings, or the recognizable shape of an outdoor baseball diamond

— *Size*: noting the relative and absolute sizes of objects is important in their identification. The scale of the image determines the absolute size of an object. As a result, it is very important to recognize the scale of the image to be analyzed.

— *Image Tone or Color*: all objects reflect or emit specific signatures of electromagnetic radiation. In most cases, related types of objects emit or reflect similar wavelengths of radiation. Also, the types of recording device and recording media produce images that are reflective of their sensitivity to particular range of radiation. As a result, the interpreter must be aware of how the object being viewed will appear on the image examined.

— *Pattern*: many objects arrange themselves in typical patterns. This is especially true of human-made phenomena. For example, orchards have a systematic arrangement imposed by a farmer, while natural vegetation usually has a random or chaotic pattern.

— *Shadow*: shadows can sometimes be used to get a different view of an object. For example, an overhead photograph of a towering smokestack or a radio transmission tower normally presents an identification problem. This difficulty can be over come by photographing these objects at sun angles that cast shadows. These shadows then display the shape of the object on the ground. Shadows can also be a problem to interpreters because they often conceal things found on the Earth's surface.

— *Texture*: imaged objects display some degree of coarseness or smoothness. This characteristic can sometimes be useful in object interpretation. For example, we would normally expect to see textural differences when comparing an area of grass with a field corn. Texture, just like object size, is directly related to the scale of the image.

Geographic Information Systems (GIS)

The advent of cheap and powerful computers over the last few decades has allowed for the development of innovative software applications for the storage, analysis, and display of geographic data. Many of these applications belong to a group of software known as Geographic Information Systems (GIS). Many definitions have been proposed for what constitutes a GIS. Each of these definitions conforms to the particular task that is being performed. Thus, the activities normally carried out on a GIS include:

— The measurement of natural and human made phenomena and processes from a spatial perspective. These measurements emphasize three types of properties commonly associated with these types of systems: elements, attributes, and relationships.

— The storage of measurements in digital form in a computer database. These measurements are often linked to features on a digital map. The features can be of three types: points, lines, or areas (polygons).

— The analysis of collected measurements to produce more data and to discover new relationships by numerically manipulating and modeling different pieces of data.

— The depiction of the measured or analyzed data in some type of display - maps, graphs, lists, or summary statistics.

The first computerized GIS began its life in 1964 as a project of the Rehabilitation and Development Agency Program within the government of Canada. The Canada Geographic Information System (CGIS) was designed to analyze Canada's national land inventory data to aid in the development of land for agriculture. The CGIS project was completed in 1971 and the software is still in use today. The CGIS project also involved a number of key innovations that have found their way into the feature set of many subsequent software developments.

From the mid-1960s to 1970s, developments in GIS were mainly occurring at government agencies and at universities. In 1964, Howard Fisher established the Harvard Lab for Computer Graphics where many of the industries early leaders studied. The Harvard Lab produced a number of mainframe GIS applications including: SYMAP (Synagraphic Mapping System), CALFORM, SYMVU, GRID, POLYVRT, and ODYSSEY. ODYSSEY was first modern vector GIS and many of its features would form the basis for future commercial applications. Automatic Mapping System was developed by the United States Central Intelligence Agency (CIA) in the late 1960s. This project then spawned the CIA's World Data Bank, a collection of coastlines, rivers, and political boundaries, and the CAM software package that created maps at different scales from this data. This development was one of the first systematic map databases.

In 1969, Jack Dangermond, who studied at the Harvard Lab for Computer Graphics, co-founded Environmental Systems Research Institute (ESRI) with his wife Laura. ESRI would become in a few years the dominate force in the GIS marketplace and create ArcInfo and ArcView software. The first conference dealing with GIS took place in 1970 and was organized by Roger Tomlinson (key individual in the development of CGIS) and Duane Marble (professor at Northwestern University and early GIS innovator). Today, numerous conferences dealing with GIS run every year attracting thousands of attendants.

In the 1980s and 1990s, many GIS applications underwent substantial evolution in terms of features and analysis power. Many of these packages were being refined by private companies who could see the future commercial potential of this software. Some of the popular commercial applications launched during this period include: ArcInfo, ArcView, MapInfo, SPANS GIS, PAMAP GIS, INTERGRAPH, and SMALLWORLD. It was also during this period that many GIS applications moved from expensive minicomputer workstations to personal computer hardware.

Components of a GIS

A Geographic Information System combines computer cartography with a database management system. Figure 8 describes some of the major components common to a GIS. This diagram suggests that a GIS consists of three subsystems: (1) an input system that allows for the collection of data to be used and analyzed for some purpose; (2) computer hardware and software systems that store the data, allow for data management and analysis, and can be used to display data manipulations on a computer monitor; (3) an output system that generates hard copy maps, images, and other types of output.

Two basic types of data are normally entered into a GIS. The first type of data consists of real world phenomena and features that have some kind of spatial dimension. Usually, these data elements are depicted mathematically in the GIS as either points, lines, or polygons that are referenced geographically (or geocoded) to some type of coordinate system. This type data is entered into the GIS by devices like scanners, digitizers, GPS, air photos, and satellite imagery. The other type of data is sometimes referred to as an attribute. Attributes are pieces of data that are connected or related to the points, lines, or polygons mapped in the GIS. This attribute data can be analyzed to determine patterns of importance. Attribute data is entered directly into a database where it is associated with element data.

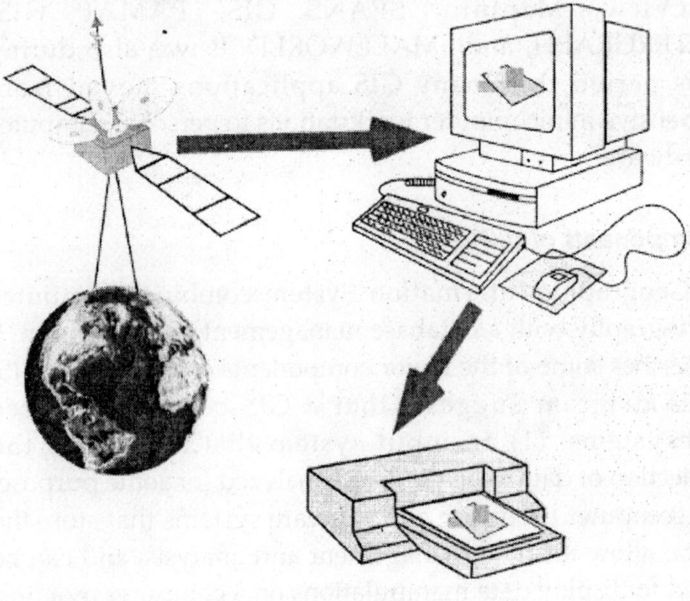

Figure 8: Three major components of a Geographic Information System.

Within the GIS database a user can enter, analyze, and manipulate data that is associated with some spatial element in the real world. The cartographic software of the GIS enables one to display the geographic information at any scale or projection and as a variety of layers which can be turned on or off. Each layer would show some different aspect of a place on the Earth. These layers could show things like a road network, topography, vegetation cover, streams and water bodies, or the distribution of annual precipitation received.